Bush & Arctic Pilot

Bush & Arctic Pilot
a pilot's story

A. R. (Al) Williams

Copyright © 1998 A.R. Williams
2022 *Reprint*

Cataloguing data available from Library and Archives Canada
978-0-88839-167-4 [2022 reprint]
978-0-88839-433-0 [1st edition]
978-0-88839-758-4[epub]

All rights reserved. No part of this publication may be reproduced, stored in a retrieval system or transmitted, in any form or by any means, electronic, mechanical, audio, photocopying, recording, or otherwise (except for copying permitted by Sections 107 and 108 of the U.S. Copyright Law and except for book reviews for the public press), without the prior written permission of Hancock House Publishers. Permissions and licensing contribute to the book industry by helping to support writers and publishers through the purchase of authorized editions and excerpts. Please visit www. accesscopyright.ca.
Illustrations and photographs are copyrighted by the author or the Publisher unless stated otherwise.

COVER DESIGN: J. Rade, M. Lamont
COVER IMAGES: istock [shaunl & ChrisBoswell]
PRODUCTION & DESIGN: J. Rade, M. Lamont
EDITOR: D. MARTENS

We acknowledge the support of the Government of Canada through the Canada Book Fund and the Canada Council for the Arts, and of the Province of British Columbia through the British Columbia Arts Council and the Book Publishing Tax Credit.

Hancock House gratefully acknowledges the Halkomelem speaking peoples whose shared and asserted traditional territories our offices reside upon, on the headwaters of the Tatalu River.

Published simultaneously in Canada and the United States by

HANCOCK HOUSE PUBLISHERS LTD.
19313 Zero Avenue, Surrey, B.C. Canada V3Z 9R9
#104-4550 Birch Bay-Lynden Rd, Blaine, WA, U.S.A. 98230-9436
(800) 938-1114   Fax (800) 983-2262
www.hancockhouse.com   info@hancockhouse.com

Contents

Preface 7
1 Early Excitement 11
2 My First Commercial Flying Position 19
3 Brief Encounter with a Legend 31
4 North of Fifty-Three 33
5 Archie and Ingy 47
6 Winter Flying 63
7 Oscar Erickson's Belly 73
8 Other Adventures 79
9 New Experiences 85
10 Aimie Lake Excitement 97
11 Wing-strut Bolt Check 103
12 Joe Bear and His Boy 107
13 Arctic Adventures 111
14 Fired 149
15 Custer Channel Wing 157
16 Looking for Work 159
17 Eldon Labe Operations 165
18 Back to the Future 183
19 Down to Earth 187
20 Northward Aviation Adventures 193
21 The Norman Wells Experience 215
22 The Yukon 219
23 Mercy Flight 227
24 Mackenzie Valley and Great Bear Lake 231
25 Yukon Territory and Home 239
Epilogue 253

Preface

Originally, this collection was intended for my grandchildren and subsequent generations of the Williams family, but when I gave copies of various stories to friends, family and acquaintances, they urged me to add to them and have them published. This was finally accomplished because, in addition to their urging, I was eventually convinced that my story is part of Canadian history that would be lost unless I wrote it down.

From June 1955 until October 1967, (including an interval in there when I operated my own electronics servicing business in Fernie, B.C.) when I was a commercial bush and arctic pilot, Canadian civil aircraft were registered as *CF-* followed by three letters. For a number of years now, Canada's aircraft registration has been changed. Today's airplanes are registered as C-F or C-G, followed by three letters. Noorduyn Norseman CF-IRE was accidentally burned at Eskimo Point (now named Arviat) on Hudson Bay when Henry Korpella was refueling her. If the aircraft had been registered today, it would have been registered C-FIRE, which to my mind would have become a self-fulfilling prophesy.

Robert H. Noorduyn, son of the designer and builder of the famous Noorduyn Norseman, once told me, "In the United States aircraft have mostly numbers in their registration, such as N7632L, whereas you Canadians call them by their first names." Keeping this in mind, I hope the reader will, throughout these pages, forgive the author for alternating between the full registration, such as CF-GTP, and a shortening to GTP, IXU, FQB, etc., thus calling them "by their first name"! I have done this so the reader will get the feeling of intimacy most pilots feel for individual aircraft. These machines do develop personalities of their own.

Still on the subject of A/C registration, a number of years ago, an Aero Commander was based at the Edmonton Industrial Airport (now called the Municipal Airport). This Aero Commander held the registration CF-SEX. I could see the reader's confusion if I had written something like, "In the minus 40 temperature, the ground crew members had SEX in the hangar to keep warm." In such a case I would most certainly have said, "had CF-SEX."

Bush and Arctic Pilot

I wish to express my thanks to my wife, Louise, for her patience in listening to my reading various passages and for her proofreading. My thanks also extends to Bruce Gowans for a good deal of input about aircraft registrations and names of persons involved in aviation and mining, especially in the Flin Flon area. Bob Lundberg is owed a heartfelt "thank you" for many fine photographs of Arctic and bush locations and aircraft mishaps. My sister-in-law Dorothy Williams, of Saskatoon, Lila Dore of Kelowna, my daughter-in-law Velma Noble of Calgary, and Angela Harrop of Naramata, B.C., also deserve honorable mention for their support and encouragement. There are many others, too numerous to mention individually, who offered vital support. I particularly wish to extend my sincere thanks to Elly Turner of the Manitoba Law Society, together with Kathy Martin of the Manitoba College of Physicians and Surgeons. These two people were instrumental in identifying Evans Premachuk, the student lawyer who successfully defended me against a charge of "Taking off from an unlighted airdrome at night." This charge was laid the morning of June 29, 1959, following an incident observed by an RCMP officer at 23:15 hours on June 28.

I must also thank my brother, Jack Williams of Saskatoon, for stories about his days as a student pilot with the Royal Canadian Air Force. His reference to old wartime pilot logbook records made possible the accuracy needed for locations, dates, aircraft identifications and names.

The tales in this book are mainly about my own experiences but I did not limit myself to my own stories exclusively. It is to be noted that there are other stories, such as Bob Lundberg's Arctic adventures, which have been included as well.

I have reconstructed conversations as accurately as possible from memory. Several of these conversations were with persons from different ethnic backgrounds, Indigenous people, Scandinavian or other groups. If any individual or group feels that I have portrayed them in a less than favorable light, I sincerely apologize. I have the highest respect for the diverse origins of the people I have included in this book.

In *Trail of the Wild Goose* (now out of print), author H.P. (Hank) Parsons used the spelling Bert Wartig for his employee. Since Hank wrote many paychecks for Bert, I believed he had spelled the name correctly.

I have since learned that this name is spelled *Warttig*. Anyone familiar with Parsons' book is hereby advised that we are referring to the same person, even though the spelling we have used is different.

I hope the reader enjoys reading this book as much as I have enjoyed living it.

Bush and Arctic Pilot

1   Early Excitement

Tiny raindrops, like thousands of glass beads, cascaded across the hood and down the right fender of my freshly polished, light green 1952 Chevrolet. As I rounded the corner from Banning Street onto Ellice Avenue, heading for Stevenson's field on the morning of November 25, 1955, lamp standards appeared to hang vertically, beneath the pavement, as the rain began to slacken. I intended to make a few touch-and-go circuits this morning, ceiling and visibility permitting.

Upon my arrival at the airfield, I would know immediately whether or not Visual Flight Rules (VFR) flight was permitted. Whenever ceiling or visibility was below VFR minimums, the control tower would switch on their rotating beacon. This would send a message to all would-be VFR pilots to avoid calling the tower.

The rain had virtually stopped when I reached the Winnipeg Flying Club parking lot; its gravel had absorbed most of the water and was nearly dry as I stepped from my car and moved to the flying club office. The inevitable group of young pilots and would-be pilots buzzed away in conversation. The smell of frying bacon, eggs and fresh coffee wafted from the kitchen into the office as I checked out Cessna 140 CF-DMQ.

Ground control cleared me to runway 34, and when I had finished my run-up I reached for the microphone.

"Winnipeg tower, DMQ ... I'm ready for takeoff."

The controller's voice sounded in my headset. "DMQ, Winnipeg tower, cleared for takeoff."

At 500 feet I was in the cloud base, and once again I called the tower, "Winnipeg tower, DMQ ... I'm just about in cloud at 500 feet, request landing clearance."

The controller's voice sounded once more. "DMQ, Winnipeg tower ... you'll be number one for approach."

I knew that the weather teletype would carry a PIREPS note. "Pilot reports ceiling 500 feet."

I had turned onto my downwind leg when a thin, metallic voice from an air force jet crackled, "Winnipeg tower, air force 321 Charlie, descending two two, out of one zero thousand, request a flat break."

This meant he was leaving ten thousand feet at twenty-two minutes past the hour. His "flat break" request was military jargon for requesting clearance from the tower to descend to 500 feet, cross Stevenson's field at 500 feet, and then line up to approach to runway 34 (this is 340 degrees, magnetic).

"Air Force 321, Winnipeg tower ... flat break is approved ... one light aircraft, on downwind leg at this time, is our only reported local traffic."

"Ahh. Roger, 321," replied the jet.

I continued my downwind leg, wondering how it felt to fly an air force jet; Winnipeg tower suddenly came on the frequency.

"321, Winnipeg tower, we have you in sight. Ahh ... cleared across east to west. Watch for the traffic."

"Ahh ... Roger ...321."

As I looked across the field to my left, I spotted that T-33 trainer, which was rapidly approaching DMQ. The controller's voice sounded panicked now; words spilled out one on top of the next.

"Three twenty-one, Winnipeg tower, watch that traffic Break left!" (He should have said, "break right.") The jet continued to bear down on me.

All sense of time vanished for me—minutes became hours, hours became days. I was transported back to my first childhood encounter with an airplane. In my mind's eye, I saw myself as a small boy playing near the gate leading to our Saskatchewan family farm.

* * *

A faint buzz grew steadily; it seemed to come from the south, although I could see no sign of the source of the sound. Then I spotted a light aircraft descending, no more than 200 feet from our gate. As the aircraft passed me, perhaps fifty feet from the ground, it suddenly began to climb again, close enough for me to see the bears in the cockpit. *Bears* in the cockpit?!

The idea of bears flying airplanes might have come from some child's book, with a picture of Daddy Bear taking his family for an airplane ride. However, I was convinced that bears flew airplanes. I felt it was a shame because, to my four-year-old brain, people should also be allowed to fly. The small, yellow high-wing aircraft continued its northern course and was soon out of sight. I made up my mind that I, too, someday would learn to fly with the bears.

Another memory flashed through my mind. When my brothers Murray and Jack joined the Royal Canadian Air Force and became flying instructors, I had learned that humans (not just bears) could fly. I longed for the day I could become a pilot.

As quickly as I had flashed back to my past, I was now back in CF-DMQ and the T-33 was closing fast; there was nothing I could do to take any evasive action, because my speed was less than a hundred miles per hour. The jet was closing in on me at more than three times that speed, seemingly destined to terminate my earthly existence.

At the last instant, the T-33 banked right to sail past my tail, close enough so the rush of the jet engine was heard over the noise from the eighty-five-horsepower continental engine that powered Cessna 140 CF-DMQ.

The T-33 turned left and headed parallel to my flight path, but a few hundred feet to my right. It turned left again, this time crossing my path, onto its base leg and then onto final approach.

I was just beginning to relax, the adrenaline flow slowing, when I was violently tossed like a toothpick. The result of the vortices generated by the jet-wake of that T-33. I was slammed down into the seat so violently that I had to pull myself up with two hands before I could see through the windscreen.

My aircraft was suddenly rolling left. I had full right aileron applied but she still wanted to roll to the left. The T-33 was cleared to land and cleared to turn right at the intersection but he turned left instead, ignoring the controller's instructions entirely.

By the time I had turned onto final approach, I had DMQ level, using some right rudder, but it was still necessary to maintain a lot of right aileron just to maintain level flight. As 1 reduced speed to normal approach speed, DMQ began rolling to the left; she came back to even keel when my speed had been increased.

"Winnipeg tower, DMQ," I said into the microphone, "I've suffered some damage here, request approach at a higher speed for controllability."

The controller came on the radio once again.

"DMQ, Winnipeg tower ... request approved...are you going to be okay?"

"DMQ ... Roger," I replied, "But she wants to roll left at normal approach speed."

I came over the fence at eighty-five miles per hour, flying DMQ to within three feet of the runway surface, reducing power. As my speed fell off, DMQ began to lower her left wing, but being only a couple of feet from the runway now, the left wheel settled onto the concrete. I held her straight, keeping her tail wheel off the surface, until our forward speed had dropped enough so she would not become airborne again; then I allowed the tail to touch down.

"DMQ, Winnipeg tower," he said, "You're cleared to the flying club area ... are you okay?"

"Winnipeg tower, DMQ, ahh ... Roger, I'm okay. Thanks." Back at the Flying Club hangar, DMQ was inspected. My left aileron was twisted badly out of shape and I was told how fortunate I had been, in view of the fate of a number of light aircraft that had begun shedding airfoils when encountering jet wakes.

Cessna DMQ was an early model of the 140. She had fabric wings and was sometimes called a Ragwing 140. Later models, such as the Flying Club's CF-GRX, had all-metal wings, the shape of which was similar to the double-taper designs that are typical of Cessna. The flat spring-steel undercarriage legs were about a half-inch thick and approximately four inches wide where they were attached to the fuselage, tapering to perhaps two inches, where axles were bolted to the legs for mounting wheels or skis.

The geometry of DMQ was such, however, that a nasty habit existed of rolling onto her nose with even moderate application of braking forces. To offset this tendency, Cessna's engineering department had designed a solution. The axles were removed from the legs and short stubs were bolted to the main legs where the original axles had been. These small stubs were fitted with the axles in such a way that the center of the axle was moved three inches forward. This solved the problem of nose-heaviness, but it introduced a new problem.

Because of the lively spring action of the undercarriage legs, any drift upon touchdown was greatly multiplied. For example, let us assume the airplane was drifting to the right just as the wheels touched the runway surface. Those undercarriage legs would twist to the left, placing more twisting forces on the legs, which would in turn cause even more twist. This was cumulative, and it made landings rather touchy if the pilot was to avoid groundlooping.

For the GRX series of 140, Cessna redesigned the undercarriage geometry to reduce the tendency toward nose-heaviness, which eliminated the stubs for repositioning the wheels, and thus, new 140s were as docile as any aircraft of their day. The redesign was the Cessna 140-A.

I loved DMQ for a few reasons: due to her reputation for groundloops, most students avoided DMQ like the plague. I could have her almost whenever I wanted to fly. Winnipeg Flying Club had lowered her rental from ten dollars per hour, which they charged for GRX, to eight dollars an hour for DMQ. When flying within my limited budget, it was a great incentive for using her instead of GRX.

Due to her twitchy handling on landing, I was convinced that if a pilot could master DMQ, he'd be able to fly anything—a fine aircraft for the learning process.

On June 3, 1955, 1 flew DMQ for the last time, taking fellow student Bill Wehrle on a half-hour local flight. This was just prior to my commercial pilot's flight test, the same day.

Air regulations inspector McMurdo came to Stevenson's field and interviewed me prior to my flight. McMurdo said me he wanted me to take off and climb to 1,500 feet, and I was then to position my aircraft above the cemetery, just north of the airport boundary, await for instructions from the control tower and to execute a spin. (He also told me that he would not be riding with me but would watch from the tower, issuing instructions through the controller.)

During the following twenty minutes, I did as 1 was told. When I was in position I executed a spin over the cemetery, and then I was instructed to close the throttle and make a landing from my present position without adding power during my landing approach.

Back on the ground again, I was naturally anxious to find out how I had done on my commercial flight test. I called the tower, asking to speak with Inspector McMurdo. At that point, the controller told me:

"Oh, McMurdo isn't here, he took off for town the moment you were airborne—McMurdo never flies with a pilot on his commercial check. I think he's afraid of flying."

The idea of executing spins over a graveyard always seemed, to me, rather ominous, but the result of the exercise was the issuance of my commercial pilot's license, WGC-7175. During flight training I'd maintained an amateur radio station under the call-sign VE4XW. When I was informed the radio amateurs at Dryden, Ontario, were putting on a mini "hamfest," I simply had to attend. I had met most of the hams personally and I had made a few on-the-air contacts with most of those remaining.

I drove to Dryden on June 8 and stayed with friends, Vic and Inez Oliver. Vic suggested the need of a seaplane endorsement, and next day he introduced me to Joe Amodeo. Joe was a certified flight instructor who also owned Piper J-3 Cub CF-GKU, on floats.

* * *

On the subject of floats, allow me to digress for a moment to let the reader know that no self-respecting bush pilot or engineer would ever refer to them as anything but floats. There was, however, a single contrary case to which I was a witness.

The de Havilland Otter is longer with respect to her ratio of airframe to float-length than is her smaller sister the Beaver, or the Noorduyn Norseman. She also has a larger side-area, which together with her length makes for more difficult docking when she is operating with a high crosswind component.

In the summer of 1953 or '54, Harry Speight was trying to get Otter CF-ODK to the Department of Lands and Forests Air Service dock on Abram Lake, near Sioux Lookout. A high crosswind component and fairly rough water made his task much more difficult than it might otherwise have been, had he been in a Beaver, Norseman or most any other float-equipped airplane.

Engineer Vaughn Downey was trying to help Harry dock CF-ODK and needed help. Downey came running into the hangar to alert other engineers to his needs. Vaughn Downey had a stutter, especially when excited, and his problem was greatly magnified on this windy day.

"The w-w-water is o-o-o-v," he said, urgency and frustration adding to his discomfort. "The w-w-water is o-o-over ... the f-f, the water is right o-o-over the f-f-f ... goddamn pontoons!" With help, Harry was able to dock his de Havilland Otter. It was only from necessity that Vaughn used the word *pontoons* rather than floats.

* * *

Amodeo had me spend an hour and a half in GKU, making simulated glassy water landings. It wasn't possible to do "the real thing" because of a slight chop at the lake's surface. He would have me approach the lake to arrive at the shore about fifteen feet above the trees; we would be traveling fifty miles per hour at that point and descending at a hundred feet per minute. Our touchdowns were accented by our floats chattering across those small wavelets, but Joe was satisfied that I had mastered the art of glassy-water techniques; the only thing left now was to experience "the real thing," as he termed it.

Later, on June 13, when Bob Peel checked me out on float-equipped Aeronca Chief CF-EVQ and Stinson 108 CF-FQB, he had me increase my approach speed to eighty miles per hour, descending 200 feet per minute to better suit the higher speed aircraft. This combination yielded very gratifying arrivals, but I had yet to execute any true glassy-water landings.

Only a few weeks later, after I had obtained my commercial pilot certificate, I was busy sending my application to various bush operators. But a newly graduated pilot is never in great demand. My experience with radio equipment was, however, about to help launch my flying career.

Bush and Arctic Pilot

2 My First Commercial Flying Position

I received an unexpected phone call from Rusty Myers, who owned and operated his own flying service out of Fort Frances, Ontario. Fort Frances is just across the river from International Falls, Minnesota. Rusty Myers' Flying Service depended on customers from both sides of the Canada/U.S. border.

Myers had a radio frequency of 5815 kHz dedicated to his flying service, but none of his radio equipment was working. He had spent more than $1,000 without results to make his communications system operational. Joe Stevens (the radio inspector from Fort William, now called Thunder Bay) had gone to visit Rusty, and he told him that unless he began using this frequency, the department would assign it to another operator.

Rusty was in a panic. "I know a man who can get your radios working," Stevens said. "But he just got his commercial pilot's license, and while he can certainly get your radios working, he is going to want to fly as well." Rusty told Joe he would call me.

Myers offered to hire me as a pilot, but only if I also promised to do the radio work. When I arrived at his base, Myers immediately instructed me to work on the radio gear. I refused to look at any radio equipment until I had flown fifty hours as a commercial pilot. My new boss gave me excuses as to why that wasn't possible—the insurance wouldn't cover me because I did not have enough hours. Only when I declared that I was going back to Winnipeg without even looking at his radio gear did he find a way to get me placed on his aircraft insurance list, and I flew about fifty hours.

I flew for Rusty Myers Flying Service during a very hot summer in 1955. Bud Kragg, Bob Peel, and a fellow known as Al Nelson were the other pilots. Of course, Rusty did his own share of flying as well.

Nelson was one of those people one might call colorful, eccentric or even strange. He wore a ten-gallon cowboy hat with bathing trunks and running shoes. He had a gunfighter's bushy moustache, and he would never load or unload his airplane. He virtually never spoke to his passengers except to say, "No, I can't go out of my way so you can see that sawmill," or any other sight his passengers requested.

In contrast to Nelson, I always wore durable but neat khaki Ontario Department of Lands and Forests uniform pants, shirt, black tie and polished black shoes. I had obtained the uniforms when I was the district radio operator at Sioux Lookout. (After leaving Lands and Forests. I removed the insignia from the uniforms and continued to wear them.) I used to load and unload my own airplane and talk to my passengers, while taking them out of my way to view landscape of interest to them.

At each day's end. Al Nelson would tell us how much money he had received in the form of tips. On the other hand, I never received a penny in tips. It was suggested to me that Nelson looked like he needed tips to allow him to buy clothes. I, on the other hand, in my spit-and-polish uniform, looked as though I was relatively prosperous.

Rusty Myers also employed a young helper named Ron Howarth, whose mom used to invite me to an occasional evening meal. An older crew member was called Jack but I can't recall his last name. Our aircraft engineer's helper was named Noel and held a private pilot license. Noel later became a missionary aero engineer with Wycliff Bible Translators.

Perhaps because this was my first real flying job, I vividly remember flying incidents.

A week after I was hired, we experienced our first relatively wind-free afternoon. I had been out on a flight. Upon my return, Rainy Lake was a glass-smooth mirror. Until then, I had never done a "for real" glassy-water landing. I naively expected it would be similar to a simulated approach.

About 600 yards to the south, we had a railroad bridge built about fifteen feet above the lake, which cut across the river mouth at an angle. I decided my approach should be over that trestle. I lined up a half-mile south of that bridge, and at eighty miles per hour, descended at 200 feet per minute, crossing the railroad at an altitude of thirty feet.

After I crossed the railroad trestle, I discovered that I had no visual reference to guide me, and I waited for my floats to come in contact with the surface of the lake. I waited. And waited. And I waited some more—surely I should be down by now. Yet I knew I was still in the air. Finally, I saw a white sheet of water flaring out to my left. I had touched down so softly that I felt absolutely no contact. When I taxied to the dock, the other pilots told me that the keels of my floats had been about six inches from the water for more than 200 yards and they had begun to think I would run out of lake before I got her down. They also told me that even after my float-keels had touched I had continued another hundred yards before I realized I was actually on the water. Finally, I could say that I had mastered glassy-water landings.

A few days later, Rusty got a booking for a flight to Flin Flon, about 700 miles away in northern Manitoba. The flight was scheduled to leave in two days' time; this was a trip Rusty didn't want to miss. I decided to get the Johnson Viking II transmitter working.

In the recent past, someone had modified this transmitter in such a way that its output power had been reduced to a very low level (this was done unintentionally, I am sure). However, the output was so low that our aircraft could only hear our calls when in visual range.

Without going into the technical details (which involved going back to original components), I was able to restore the full 100-watt output from the Johnson Viking II transmitter.

In Rusty's Norseman CF-OBD, I discovered that his antenna was electrically too long for our frequency of 5815 kHz, and I found a way to shorten it electrically without making a physical change in its length, which would have called for paper reports to have been generated to satisfy the Department of Transport.

Through these two actions, I had restored the operation of both. Rusty Myers would have good communications on his trip to Flin Flon.

The following morning, when Rusty was en route, he called every ten minutes to make sure the radio was still operating. Communications remained excellent for the entire trip. Rusty was beside himself. Finally he called to announce his arrival. "Fort Frances, OBD, we're landing Flin Flon in five minutes." I acknowledged his landing report and said we'd see him later. Just more than five minutes later, Rusty's weak signal was heard. "Fort Frances, OBD ... on the water at Flin Flon."

"OBD," I said, "Check, you on the water at Flin Flon." That 700 miles or more from Fort Frances to Flin Flon was a pleasant surprise to Rusty when compared with his radio's former range of five or six miles.

As soon as Myers returned to Fort Frances, he put pressure on me to get the other aircraft radios working. I refused to look at any other radios until I had logged another hundred flying hours, at which time I repaired the rest of them.

We settled into a routine of fishermen drop-offs and pickups, until Chalky Martin taxied his Cessna 170 to our seaplane dock at 06:00 hours on a Sunday morning in late August. While Chalky was an American and his Cessna carried American registration, he held a Canadian private pilot's certificate, which allowed him limited commercial use of his aircraft. He could supply his own camp with groceries and other necessities, but he was not approved to carry paying passengers.

Chalky reported that Stinson 108-2 Voyager, which was flown by Vern Jones, had met with an accident. During takeoff, he said, the float bottoms had apparently failed and the aircraft had sunk in about twenty-two feet of water. Chalky reported that everyone on board had been saved because at the time of the accident people on shore had launched a rescue boat before the aircraft had gone under.

Although Jones was employed by Rainy Lake Airways, there was no salvage help available there because the airways had only two aircraft. Metro Kirby was away with Norseman CF-ECC on a three-day charter.

Rusty Myers assured Chalky Martin that a salvage operation would be mounted immediately. Because this was my first season of seaplane flying, I had never been involved with a rescue operation. I asked if I might come along. In addition to several ropes, we took a surplus U.S. Navy rubber raft, a small outboard boat motor, several long two-by-six planks, forty-five imperial gallons of 80/87 aviation gasoline and five gallons of engine oil. The younger members of the crew gathered their swim trunks and other necessary personal gear. Within minutes, we were airborne in Norseman CF-OBD, with Myers at the controls.

As we circled the lake, we could see the Stinson. She was near the middle of the lake, mainly submerged, with about four inches of the lower section of her rudder out of the water. She was resting in a semi-inverted position, on a tripod consisting of her floats and her engine cowling. Judging the angle of the float bows to the cowling, it seemed Chalky had been fairly close in his estimate of the depth of the water. Rusty judged it to be about twenty feet.

Rusty Myers taxied OBD toward the submerged aircraft and expertly worked the mixture control so we stopped exactly in position above the submerged Stinson 108-2. Clear water offered us good visibility. We could see the problem. Both float bottoms had obviously collapsed inward; at the time the Stinson was traveling about fifty-five miles per hour. The force of water rushing into the floats had forced excess water to rush back out, causing the float bottoms to rip open outward. Jagged metal extended out from the float bottoms as if a bomb had exploded inside them.

The men quickly donned their swimming trunks, and in a moment were down at the front spreader bar of the Stinson, tying ropes to both ends at the float attachment points. Jack and Ron quickly secured the ropes to our float cleats. These cleats are often referred to as *bollards*, and, indeed, they do serve the purpose, but bollards are physically different. I could not understand how we planned on raising that aircraft from twenty feet of water using just a pair of ropes.

Ron and Jack made another dive and secured two additional ropes to the spreader bars. Next, they brought the two, newly attached, ropes to the surface (after passing them under the leading edges of the Stinson's wings). This was to ensure the Stinson would assume its normal horizontal position—no point in bringing her to the surface inverted. Myers had me help him position three of the planks across the Norseman floats to form a platform beneath the propeller. The two newly attached ropes were wound around the propeller hub, which, together with the propeller blades, acted as our windlass; we began the slow job of raising the Stinson out of her watery grave.

After only three or four turns, the men secured the first ropes to the cleats (also passed around the leading edges). We would back off our windlass to get a new purchase for three or four more turns.

In the end, we had the leading edges of the Stinson's wings level with the water; Ron and Jack submerged a deflated rubber raft and positioned it beneath the keels of her floats, but only after doing a repair on the jagged float bottoms to prevent damage to any part of the raft. They then partially inflated the raft to buoy the plane's wings up, near the surface of the lake.

The wings were saturated with water, and if the raft had been inflated any further, the underside fabric would have been damaged. The two men swam around the perimeter of the wings, making small slits at the trailing edge to let the water out slowly. The wings began to ride higher in the water, until, at last, a good two inches of clearance could be seen.

Ron and Jack dived underwater again, inflating the rubber raft even further until the Stinson's belly was just above water level. This was also done slowly, which allowed the water trapped inside the fuselage to escape. A sudden wind sprang up, blowing from the west, and we (Stinson, Norseman and all) began drifting toward the rocky shore on the eastern end of the lake.

We could not use the Pratt & Whitney R-1340 engine on OBD to help move against the wind, since it was still tied up as a support for the Stinson. Paddling proved ineffective. Rusty and I placed one more plank across the decks of OBD's floats; this time it was placed just aft the rear spreader bar, and the outboard motor was secured to this plank.

The outboard lacked the power to move us against the west wind, but with two men paddling we were able to hold our position. Within twenty minutes the wind died down; we were able to proceed, very slowly, toward the camp at the west end of the lake, using the power of the outboard, while two of us stood on opposite ends of the plank to balance it.

Our headway was slow. We steered mainly with water rudders and also paddles. Once we had the Stinson on shore (I did not record her registration), the rubber raft was fully inflated. Blocks of wood were placed under spreader bars so that the keels of her floats were several inches above the shallow water at the shore of the lake.

The fuel lines were disconnected from the tanks, and the tanks were drained. The engine oil was drained and the system flushed. Magnetos, which provide spark for ignition, were removed and taken into the main building of the camp on Turtle Lake; they were placed on an oil stove to dry, and the carburetor was placed near that stove, after it had been drained and flushed with gasoline, kept far enough away so it presented no fire hazard.

In the meantime, Ron and Jack began to fashion makeshift float bottoms, using sheet aluminum we had brought with us for the job. This was not intended as a permanent repair but as a temporary measure for the flight to proper repair facilities. It was decided the missing right door could be ignored (one of the passengers had yanked the door off to make a hurried exit when the airplane was sinking). The next problem was the missing windshield. Jones decided that a towel wrapped around his head and a pair of sunglasses would suffice for the ferry trip back to Fort Frances.

After the evening meal, Howarth and I sat on the dock, smoking and watching a gibbous moon floating in the southern sky. Ron and I were good friends; he had sometimes ridden with me in our company Stinson Voyager CF-FQB, and we had shared evening meals at his mother's house. We discussed the happenings of the day, the relative merits of our Voyager with a 165 Franklin engine, compared with Vern's machine and its retrofit 190 Lycoming engine. Ron commented on the extra speed delivered by the greater power and expressed the hope Vern would throttle it back (to keep from being blown out of the cabin). I exclaimed, with my youthful inexperience, that the salvage of the Stinson was miraculous.

We sat in silence for a time, watching the occasional glow from a firefly near shore. Out on the darkened lake, beyond tree and shore, a fish would break the surface, with ever-widening circles of ripples that eventually popped up beneath the float bows of OBD. Somewhere, off in the distance, a loon piped up with its haunting cry. We both agreed that life was wonderful.

Next morning after breakfast, a bustle of activity took place. The carburetor was examined and reinstalled in the Stinson. Magnetos were tested and put back into the engine. Engine oil was replaced and fuel lines were reconnected after fresh gasoline had flushed the tanks. The fuel tanks themselves were filled with about a half-hour reserve for the flight home. Ron and Jack had done their best under trying circumstances, but they feared some hefty leaks might still cause a problem: the makeshift float bottoms were fastened in place with only half the original number of self-tapping screws. We had used spruce gum as a makeshift gasket in an attempt to make those float bottoms as seaworthy as possible.

Vern did his initial run-up while resting on the sand bottom of the shore next to the dock; ropes were tied between rear float cleats and shoreline trees. When engine operation seemed as steady as one could expect, we all gave the plane a push. Wearing his sunglasses and towel, Vern taxied out on the surface.

When he was ready for his takeoff run, Vern stepped from the cabin to the left float and pumped out a great deal of water. He went to the right float and repeated his pumping exercise. Again, a sizeable quantity of water was pumped from starboard. Once again he pumped the port float, but this time not much water was removed. At this point Jones raised his water rudders and applied full power. In a remarkably short distance, he was airborne. We climbed aboard CF-OBD and in short order, we, too, were on our way back to Fort Frances. Vern landed without further difficulties at the Rainy Lake Airways base.

A few days later, I stopped by to see him. Following my single visit with Vern Jones and Metro Kirby, we went our separate ways and had only the occasional chance meeting.

Just once after the salvage job, Metro Kirby invited me along in his Norseman CF-ECC. He gave me a valuable lesson in the proper way to climb and yet retain good cooling of this aircraft's engine. He would climb at a hundred miles per hour, whereas other pilots would climb at only eighty. Cylinder head and oil temperatures would usually climb into the upper end of the green arc of their respective gauges at eighty miles per hour, while at a hundred miles per hour they would stay well inside the green ranges.

Metro Kirby showed me that a Norseman climbed at the same rate at a hundred miles per hour as it did when he allowed it to wallow along at eighty.

Visibility improved as well—although forward visibility was never a feature the Norseman could boast of. I was to be the beneficiary of this lesson later in my flying career.

Not long after Rusty Myers' flight to Flin Flon and before I restored the radios in the rest of the fleet, I had an interesting experience. Six weeks had passed since we had seen any clouds. Each day the sun beat down on us and temperatures soared; we looked forward to evening when things cooled down. In late afternoon the sun was low in the west; if it was necessary to fly westerly, pilots had to keep their eyes at least partly closed against bright sun rays.

One day I had to fly the circuit judge to a settlement east of Fort Frances. Some of the smaller settlements were accessible by road, others only by water or air. Judges were too busy to travel by either road or water to reach these settlements, so they chartered an aircraft. This particular trip took place in late afternoon, when the sun was beginning to sink toward the western horizon.

On my return to Fort Frances the cockpit was hot, and the late afternoon sun turned it into an oven. My eyes were half closed against the sun's brightness. I calculated my time of arrival at Fort Frances to be about sunset. Suddenly, the sun was gone and darkness was all about me. There were no familiar sights outside the cockpit window. I looked down to see if a familiar lake was visible; very few lakes were to be seen. I could see, dimly, the outline of fields. Here and there, yard lights could be seen in farmyards. I had drifted off to sleep and I was flying on a southwesterly heading somewhere in the northeast corner of Minnesota—likely southeast of Birchdale.

I executed a rate one turn to the left (this is a fairly gentle turn where the nose of the aircraft comes around the horizon at a rate of three degrees per second). Sixty seconds later, when I had turned 180 degrees, I could see a cluster of lights near the eastern horizon that I recognized as Fort Frances.

Over Fort Frances, I lined up to land, passing over the railroad bridge to guide me as to height. My touchdown was as gentle as I could have wanted. As I approached the dock, I reached for the mixture control to stop the 165-horsepower Franklin engine in Stinson CF-FQB, when it died of its own volition. I had run out of gasoline in exactly the right spot needed to reach the dock!

One day, during mid-August 1955, 1 had the opportunity to go along to Fort William with Bob Peel. We took our company Gullwing Stinson and departed Rainy Lake for Lake Superior. I will always remember how serious Bob Peel remained on this flight. Besides his riding with me during my checkout ride in the Aeronca Chief and Stinson 108-2 Voyager, this was the only time I spent in the air with him.

Bud Kragg, on the other hand, was the picture of complete relaxation whenever he flew Mark VI Norseman CF-GTN; the throttle lever was in the center of the throttle quadrant, the mixture on the right and the propeller control to the left. Bud would set the mixture control to full rich and advance the throttle to read thirty-six inches of manifold pressure, with the propeller lever set for full fine pitch at 2,250 rpm for our takeoff.

The moment our floats left the surface of the water, Bud wrapped his fingers around the throttle and propeller control levers and reduced the settings of both simultaneously for climb power of thirty inches of manifold pressure at 2,000 rpm. One could feel the surge of thrust as old CF-GTN leapt forward.

Whenever he landed back at our base, if the wind was in the east, Bud would approach, aiming for a patch of water just south of our dock system. He would maintain ninety miles per hour and fly within a few feet of the grass in the pasture that stretched parallel to the trail leading from the main road to our base. The pasture began up the hill beside the road, and it ended near the southwest corner of our office. Bud Kragg would wait until he was very close to the building, then chop his throttle, with the resultant backfire making a pop-pop-pop-bang noise, almost enough to wake the dead. Nobody ever mentioned the noise, probably because an east wind was quite rare and there were few such landings.

One day, Bud Kragg was dropping some fishermen off at Canoe Lake. He commenced a routine "dragging" of the lake, to search for hazards to landing, such as submerged rocks, sandbars and deadheads (logs which have become partly waterlogged with one end sunk to the bottom while the other end still floats near the surface). He failed to see a deadhead directly in his takeoff path.

The deadhead was lying parallel with the wind, pointing directly toward Bud's takeoff path. Just when GTN was at takeoff speed, the floating log end smacked the bow of his left float and Bud was literally pole-vaulted into the air. The collision was so violent that his left float was bent into a banana shape; a great rip appeared in the step compartment; the bow was pushed up and slightly to the left, so that only two inches of clearance remained between his propeller tips and the damaged float.

Bud flew back to Fort Frances. Since this was prior to my repairing his radio, he flew low and "surged" his propeller to alert us that he had a problem. Bud Kragg lined up for his landing, aiming for shore just a few yards south of our dock system. He landed GTN on her right float and taxied on the step until he was close in. He chopped power as he approached and allowed his left float to touch just as it seemed inevitable that GTN would hit the dock.

Bud was much too skillful for that to happen; as his left float came into contact with the water, GTN slewed around to her left, making nearly 300 degrees as she staggered to her left. Her left float settled into three feet of water. The skillful pilot opened the cockpit door, and, standing in the doorway, he placed his left foot on the near vertical bow of the left float. He calmly announced that he had "had a little problem."

I filed Kragg's experience away in my memory for future use. Never again would I do just a cursory inspection of a proposed landing site. I always carefully scrutinized every landing path from the air before putting an aircraft onto that stretch of water. If I found I had to land on water that was black from muskeg runoff or from coffee-colored silt, I would slowly taxi the length of the takeoff run. This approach took a bit longer but provided me with a margin of safety I might not otherwise have had. 1 was learning my craft from observing the professionals.

A few mornings later, I flew CF-FQB to Shipiro's camp, to the north, on Sphene Lake. As I opened the throttle for takeoff, I had to apply carburetor heat due to high humidity. On that takeoff run, it was necessary to shut off carburetor heat, since I needed the minor increase in power I gained by not using it.

In those days I just accepted the power decrease when carburetor heat was used, without any thought as to why that should be. I believe it occurs because heat is introduced. Hot air expands and the mixture of fuel and air becomes richer, whereas normal cold air contracts to become more dense and the mixture remains ideal for proper combustion.

As I approached the shore, about fifty feet above the trees, the engine quit on me—carburetor ice. In most other aircraft I have flown, when carburetor ice was encountered the engine would miss and perhaps backfire, but the Franklin engine just quit cold.

Two tall spruce trees stood twenty to thirty feet above the rest of the forest; I was sinking toward their tops—I couldn't get over them and I couldn't fly around them. I aimed for the center, hoping I could hit both trees simultaneously. I did. About three feet of wingtip remained in the upper branches of each spruce, and because I was now traveling nearly ninety miles per hour, I was able to put the aircraft floats on the tops of the forest as though that was exactly where I had intended to land. I learned this from James Kirk. Jim flew for Lands and Forests up at Pickle Lake; he put five aircraft into the bush over a thirty-year period. He had never been seriously injured during those forced landings. My own landing was not considered very serious, as it caused only about $7,000 damage.

When I left Fort Frances on August 29, I had something like 330 flying hours under my belt; Rusty Myers' Flying Service had an air-to-ground radio system that was second to none. There was no further difficulty getting insurance coverage, because the accident investigation had exonerated me of any blame, since circumstances had forced me to turn off carburetor heat on a Franklin engine that was almost guaranteed to quit when encountering a case of carburetor ice. Consideration was also given to the fact I had not been advised of the Franklin engine's idiosyncrasies. The ruling also found that I was not to blame under the circumstances of that particular takeoff run.

I learned a host of valuable flying skills and many other interesting things in the beautiful country surrounding Fort Frances. I have fond memories of Bud Kragg and my other fellow employees. I never forgot my first experience with aircraft salvage operations.

3 Brief Encounter with a Legend

When I returned to Winnipeg in September 1955, I waited for responses to my applications for bush flying. One day, I received a call from the legendary Tom Lamb. Tom Lamb had virtually invented flying in Manitoba, and he was the first to begin bush flying in northern Manitoba. He was also in the Arctic years before it became routine. He truly was a legend in bush flying. Lamb had seven sons, and they continued the business after his passing.

Tom owned and operated Lamb Airways (later Lamb Enterprises) at The Pas, Manitoba. He had just purchased a factory-new Cessna 180 CF-IRP. Lamb asked me to go to Brandon to pick up his new aircraft at Brandon Airport from Ed Magill. Lamb had traded an older airplane for IRP. Magill and Lamb agreed that Tom would operate both aircraft until the old one was sold, or at least until it had drawn serious interest.

When I delivered this new aircraft to The Pas, it was my first visit back since I had been employed as operator/agent for Canadian Pacific Air Lines at the airport. Now I hoped for a few hours flying from The Pas as pilot-in-command. I was to be disappointed. I was with Lamb only two days before the old aircraft was sold. The buyer arrived at The Pas onboard TransAir DC-3 service. I had to return to Winnipeg as a passenger on the DC-3, missing the chance to fly the old Cessna out.

I did learn something interesting, however. Lamb told me he had completed only Grade 1 in school. "I'd have got Grade 2, as well," he told me. "But the family moved and there was no school." Tom's mother homeschooled him but Grade 1 was the end of his formal education.

Tom Lamb spoke Cree fluently and would often converse with a Cree person in the language while someone who spoke only English stood by. Tom would talk freely to both the Cree and the English speaker in their own language as though the third person wasn't there. He meant no disrespect. This was merely a habit gained over many years of using the two languages.

Tom's father had established a trading post on Moose Lake prior to the 1920s. Most of the Lamb family's income was derived from trading muskrat furs. When a series of dry spells occurred in the late '20s or early '30s, a system of canals was constructed to preserve the water habitat in which the muskrat population could survive and multiply.

Despite his lack of formal education, Tom Lamb became a commercial pilot. In his era, a commercial pilot license required only ten hours.

Later on, when an examination was required for aircraft maintenance engineer papers, Lamb studied for and passed the government exam. When it became law that a technician had to be licensed in order to perform maintenance work on aeronautical radio, Tom Lamb studied and wrote an exam which granted him that ticket as well.

Many years later, about the mid-1940s, Tom became a cattle rancher at Moose Lake. As no veterinarian was available, Lamb studied animal science and was granted a provincial veterinarian certificate. He was permitted to attend to the health of his animals. Lamb was truly an eager, lifelong student who greatly valued both practical and scholastic education—a legend in his own time.

4 North of Fifty-Three

Back in Winnipeg in spring 1956, I sent more applications to various bush operators in hope of finding a good flying job. I had held my commercial pilot certificate for more than a year but still had only limited experience of 307 hours and fifty minutes in my chosen profession.

* * *

At this point, I must take the reader back to Thanksgiving 1949. I had been employed by Canadian Pacific Air Lines since June of that year. And after trial-by-fire lessons in dispatch work (following the hospitalization of the regular dispatcher), I had been transferred to Prince Albert, Saskatchewan, to fill the positions of radiotelegraph point-to-point operator/agent. I had decided to hitchhike home to Lac Vert to spend the holiday weekend with my family.

I had only gone as far as St. Louis, about twenty-seven miles south of Prince Albert, when I decided the rides were too few and far between, considering the cold weather, to continue. Just at that point I saw a Saskatchewan Government Transportation bus approaching from the south. I flagged the driver and hopped aboard for the short trip back to my warm room at the Prince Albert airport.

I walked down the aisle and took a seat across from an attractive young lady.

"Where did this bus come from?" I asked.

"From Saskatoon," came the reply.

Never being one to allow an opportunity to drift by, I introduced myself. She replied that her name was Louise Rollefson and that she was on her way to visit her sister in Prince Albert.

We chatted for a while and she mentioned she worked in a bank in Saskatoon, although we never got around to discussing the name of the bank. And I didn't ask for her address or phone number.

When we arrived, Louise needed a nickel to call her relatives. They had been expecting her to arrive on the train, but at the last minute she had decided to take the bus instead. I lent her the coin, and a few minutes later her brother-in-law Earl bounded through the door and from fifty feet away shouted, "What the hell are you doing here?"

Louise was embarrassed. But I thought, "This is my kind of guy."

Within a week I found myself writing a letter to Louise, beginning by stating, "I don't know why I am writing this letter." It seemed to be fate. Louise had taken the bus instead of the train, and I had decided to jump on that bus.

Although I didn't know which bank Louise worked in and I had changed her name from Rollefson to Olafson, care of the Canadian Bank of Commerce (a guess), the letter was delivered in due course. Arrangements were made for me to visit her in Saskatoon.

We married in Saskatoon on March 20, 1950.

* * *

Fortunately, that spring of 1956, Louise was employed by the Canadian Bank of Commerce in Winnipeg. She could see us through our money shortfall.

In early June, I received a call about an opening for a pilot at Flin Flon, Manitoba. I drove our 1955 green Ford to Flin Flon, arriving at Schist Lake Friday morning, June 15, 1956. The first thing Hank Parsons said to me when I arrived at Parsons Airways Northern Ltd. seaplane base was, "Have you ever had an accident?" 1 told him about the carburetor ice, treetop landing in Stinson CF-FQB. "Damn good thing you told me," he said. "If you hadn't and I found out about it, I'd fire you on the spot."

He had some bad news for me as well. "I don't have enough money in the bank to meet the next payroll. If that is a problem for you, then you had better get back in your car and go home." I assured Hank that I could survive for a time without a paycheck—just so it wasn't too long.

To soften the payroll blow, Parsons told me that he would let me stay in their family cabin, overlooking Schist Lake from a wooded knoll about thirty yards south of the base. I hadn't been there long when Louie and Dot Mayers asked if I would like to move into their house. A room in their home vacated by a Parsons pilot (Bert Warttig) was available for the remainder of the summer. I was told only that Warttig was in

Winnipeg for some undefined period.

Louie Mayers was in the commercial fishing business; the slab walls of his fish-packing house stood about thirty yards north of our seaplane base, on the north side of Northwest Arm. His fish house was filled with sawdust and blocks of ice gathered in winter and used to pack fish for shipment to market. Mayers owned a Bombardier snowmobile, on which he made his winter rounds. One day Jimmy Hoglander noticed that the machine was not parked in its usual spot. "What's gone wrong with your Bombardier?" Jim asked him. In a heavy German accent, Louie replied, "My Bombardier is stuck. It is stuck on a stump," he said. From that day on, Jimmy or Ernie would ask him some similar question, pertaining to the Bombardier, and always received a similar answer, "The Bombardier is still stuck on a stump."

For all of Louie's reluctance to speak freely, his wife, Dot, was just the opposite. Her lilting Irish accent made her a delight to listen to. Dot was, however, notorious for gossiping about the comings and goings of neighbors. She once told me, "People say that I'm a gossip, but I'm not. I only tell them what they want to hear." I suppose there was some truth in what Dot told me.

Dot and Louie had courted under somewhat unusual circumstances. Dot's former occupation was as a lady of the evening in a local bawdy house. Louie was a regular visitor to the house and always asked for Dot's services, to the exclusion of all the other working girls. Eventually, Louie asked her if she would do him the honor of becoming his wife. Dot accepted.

During my stay with the Mayers, I awoke one morning to find traces of blood on my bed sheets. I puzzled over the blood stains until the morning I was awakened by a pain on my left side. I searched around and discovered a broken spring poking up through a hole in the mattress. I was a bit heavier than Bert, and my weight had forced that broken spring up through the hole to scratch me. As soon as I brought the matter to Dot's attention, she replaced my mattress with a brand new one. Dot Mayers was a kind soul and wanted nothing short of the best for anybody with whom she came in contact.

Jimmy Hoglander and Ernie Day were the company helpers. Day was a character. He liked his liquor, and one day, although he was a tireless worker when on the job, he was three hours late for work. Hank Parsons had spoken to him about this problem several times. On this day, Parsons had run out of patience. "Ernie, you're fired," he stated. Parsons hated to fire any employee, but Ernie had been warned too many times. An hour later, Parsons looked out the window and Ernie was up on the wing of Cessna CF-IXU, filling the left wing-tank. "Didn't I fire you?" Hank asked.

"Well," said Ernie, "I didn't have nothin' else to do, so I decided I would gas up your airplane for you." He was rehired on the spot.

On another day, when Ernie had taken a round or two out of John Barleycorn, he came to work late. Once again, Parsons was going to fire him. When Ernie pushed open the office door and tossed in his hat, Parsons relented. The comical action saved Day from ending up on the unemployment lines for yet another day.

* * *

One day Bert Warttig told me about his accident in Hank's Cessna 195 CF-FLN. Warttig encountered heavy fog building over a goodly portion of the Herb Lake and Snow Lake area and attempted to land in rapidly deteriorating visibility. He had begun a turn and the next thing he remembered was waking up in a Winnipeg hospital.

I do not have the date of the accident, but I know it was shortly before Hank Parsons hired me as a pilot to fill a vacant position. Bert told me he went into the hospital at 180 pounds and at one point was down to 128 pounds. Bert was lucky to have survived.

One passenger had been killed in the Cessna 195 accident, which caused Parsons a good deal of mental anguish. Parsons was upset because of the individual's death, was concerned about the company's reputation and the fact that the aircraft had no damage insurance. There was no blame attached to Bert; it was just one of those unfortunate happenings over which he had no control. Hank Parsons was also in the area when fog began developing. Parsons was fortunate enough to be quite near his destination at the time and was able to land safely.

* * *

A. R. (Al) Williams

Speaking of Bert Warttig, I must relate this little incident. I have never been a drinking man—a fact that amazed some pilots. One evening, a number of us were visiting Parsons' family cabin, and a good deal of drinking was going on. Somewhere along the line, somebody asked if anyone would like another drink. Bert Warttig spoke up saying, "Don't give Al any more—he's schartin' to look blurred!"

* * *

Bert's accident had left us with only three aircraft. We had Cessna 170 CF-HIY, a Stinson SR-9 Reliant CF-BGS and a Cessna 180, CF-IXU. Hank had just acquired IXU. It was the first new aircraft he had ever owned. Naturally, it was his pride and joy. Parsons was the only pilot who was allowed to fly that shiny new company flagship. He had flown to Winnipeg, just after my arrival at Flin Flon, and purchased the aircraft in spite of having only $1,000 in his personal account. The purchase agreement was written on a serviette in the Marlborough Hotel coffee shop. Hank Parsons went into debt for $23,000 that day and flew the new airplane directly back to Flin Flon.

My boss was very protective of his new Cessna 180. The day following his arrival with the new aircraft, Parsons walked along the dock where Ernie Day and Jimmy Hoglander were in the process of loading IXU. "I don't want you pot-lickers putting too much of a load in that airplane," he said. "Not to worry, Hank," said Ernie Day. "It's all heavy stuff; there's no bulk here." The statement satisfied Parsons and he went back to the office. Just as his hand came to rest on the door handle, he stopped, looked back at IXU and shook his head. It had taken that long for him to recognize what Ernie had said to him.

At this time, Hank and I were doing all the company flying. I was given both Cessna 170 CF-HIY and Stinson SR-9 CF-BGS while he flew his new IXU. Between us we kept all three aircraft busy. Business was brisk that summer, with Parson's new Cessna 180 particularly in demand by customers. I recall spending 208 hours in the air during one thirty-day period that summer. Such a record could probably not occur today, with strict enforcement of the number of flying hours.

July was rapidly approaching, and we had been flying steadily every day for a few weeks. Hank told us that since we had no serious flying scheduled for July 1, we would take our aircraft to Cranberry Portage for a day of relaxation. However, we would also do a bit of barnstorming and take a few passengers for sightseeing rides—just enough to defray the cost of our flying trip.

Parsons used Stinson SR-9 CF-BGS while I flew Cessna 170 CF-HIY; we took the whole crew along, including Ernie Day and Jimmy Hoglander. Considering that Cranberry Portage was remote bush country, I was surprised at the number of residents who had never taken an airplane ride. We flew nearly as much on this non-flying day as we would normally on a working day. Parsons was overjoyed, of course, because this gave his business a shot in the arm and helped him meet the next payroll. We returned to Flin Flon that evening as tired as we would have been on any normal bush flying day.

While some customers chartered CF-BGS, our Gullwing Stinson SR-9 (this aircraft was much larger), her charter costs were considerably higher per mile than they were for the Cessna 180. We were hauling 800 pounds of payload in CF-IXU. But it was not possible to coax the 145-horsepower engine in Cessna 170 CF-HIY to drag that amount of load onto the step.

Our problem was solved by Gordon Mitchell. Gordon's primary business was Mercury outboard motors, but part of his income was generated from being a licensed (and very proficient) airframe and aero engine engineer. Between sales and service of marine engines, together with aircraft work, he kept busy. A few itinerant aircraft needed his attention, and he also had a couple of privately owned aircraft that were often in need of routine maintenance.

One day, with a good stiff south wind blowing, I was just able to stagger CF-HIY onto the step with a very substantial load on board. For the uninitiated, a seaplane float has a large floatation section from the bow to a bit past the midpoint toward the stern, and at that demarcation point there is an abrupt rise in the keel. Forward of that point is called the step compartment (floats are hollow and are compartmentalized).

While the aircraft is at rest on the water, the entire float provides floatation, but during the takeoff run it is necessary to get the tail end (the heel) clear of the water. When this happens, the step compartment is the main part of the float in contact with the water. The art of getting "on the step" must be used in order to reduce the water-to-float friction and accelerate to takeoff speed.

Gordon had noticed the difficulty that our underpowered Cessna 170 experienced in getting on the step. He made a sound suggestion.

"Why don't we try towing CF-HIY onto the step using my new Mercury mark sixty and speedboat combination?"

Parsons was a little dubious. "What happens," he asked, "if the rope gets tangled in the bracing wires around the floats?"

"In such a case," I said, "I'll just pull the throttle, stop and get it untangled. I think it's a good idea and it will get all those requests for IXU off of your back. I could go out of here with 800 pounds, just like you do with IXU. HIY can handle that kind of a load in the air. It is just in getting her on the step where we need help."

In the end, Parsons acquiesced. We used Mitchell's strategy whenever we had a larger than normal load that needed a bit of help getting on the step. Mitchell would approach my left float and place a rope, with a metal thimble in its end, around the rear point of the front port float cleat. Then he would slowly work his speed boat out in front and to my left, until the rope was beginning to tighten. When we were lined up for takeoff, he would raise his right hand, which I would accept as a signal that he was about to open the throttle on his outboard. I would place my hand on the throttle, and as his hand dropped, we would both gently advance to full power.

With an added sixty horsepower, HIY would climb smartly onto the step and Mitchell would reduce power a bit so I was catching up to him. The rope would go slack, the pull of the water would slip the thimble off the back of the cleat, and I was on my way. There was no difficulty in getting our Cessna 170 into the air once we had gotten her on the step. Although climb performance was not as spectacular as one might have wished for, it was adequate, and customers were glad to have Cessna 180 loads carried for the reduced price of the Cessna 170.

The 230 horsepower of our Cessna 180 was the main reason for a forty-five cents per mile charter rate, while the 145-horsepower Cessna 170 chartered for thirty-five cents per mile. The customer was therefore very pleased to have his load taken in the 170, although we charged him a five-dollar fee for Gordon's tow in addition to the rate for their charter.

It was still an excellent deal in comparison to Stinson CF-BGS's charter rate of sixty-five cents per mile.

A couple of months after Gordon solved our step problem, I was returning from a trip, and as I taxied back to our base I saw Gordon standing on Mitchell's dock with a serious-looking man I didn't recognize. The pair seemed to study CF-HIY as I taxied by. When I had tied up and entered the office, Ernie Day was just ending a telephone conversation.

"You have to take HIY over to Mitchell's dock right away," he told me. "Air regulations inspector wants you and HIY there right now."

I couldn't for the life of me understand why an air regulations inspector would want to see me; more puzzling was why he would ask to see CF-HIY. In any case, I taxied to Mitchell's dock and tied my aircraft to it.

Gordon introduced me to the stranger, whose name I have probably forgotten intentionally. He held in his hand a very clear color photo, which was actually the cover for a Mercury Outboard Motor Company sales brochure. It had never occurred to us until that moment that our towing operations might cause us trouble from air regulations. Somebody had taken a picture of us in the process of hauling CF-HIY onto the step. Mercury had no doubt paid handsomely for such a clear, detailed picture.

The inspector held the picture out for me to examine. "I suppose you can recognize this picture, can you?" he asked sternly. I said that the picture had never come to my attention before now, but yes, I could recognize it. At this point, I was shaking in my boots, because, while a portion of Gordon's income was earned from the outboard motor business, my entire income was dependent on retaining my pilot's licence. At that moment, Gordon and I contemplated futures outside the aviation industry.

"I'd like to witness a demonstration of this unusual coupling between aircraft and boat," he said.

"Oh boy," I thought.

We went back to the dock, and I untied HIY while Gordon Mitchell readied his speedboat. The inspector told me that he would ride in the left rear seat, directly behind me, so he could watch as the rope and thimble were used to pull us along. The inspector appeared dubious about the safety of the entire procedure. I tried to be upbeat in my responses, but he certainly did not seem to share my confidence.

Gordon slowly moved his speedboat out to our left and gently brought the rope taut. Mitchell raised a hand and I placed mine on the throttle, ready for our boost onto the step. We had never done this with only one passenger on board, and I was unsure just exactly what to expect. Mitchell's hand dropped, and we opened our respective throttles in a slow and deliberate way. CF-HIY rose onto the step just as had occurred with larger loads; the only difference was that we now used considerably less time and distance.

Gordon reduced his power setting a bit, and as the rope became slack, the pull of the water on the rope smoothly slid the thimble from the left float cleat, just as it had always done. Once airborne, my passenger asked me to go around, land and dock at Mitchell's again. Those instructions were the only words he spoke while on board my aircraft. At Gordon's dock again, the inspector deplaned. He said he wanted to watch this operation once more, but this time he would do so from the speedboat.

The procedure again went smoothly, but we used even less distance and time. Once I was in the air, I completed a short circuit and made my landing. As I turned around, the inspector waved me to Mitchell's dock once more.

When I had docked, we all went into Gordon's office, which had a very large picture window facing his dock system. The inspector stood near the window, gazing at HIY and Gordon's speedboat with his back to us; Gordon and I stood at his counter. A week seemed to go by as I stood there with my knees becoming weaker with each passing second. My heart was thumping loud enough so that I was sure it could be heard across the room. Gordon's big office clock was ticking loudly, measuring out our future in aviation. The inspector finally spoke, much louder than was necessary, it seemed to me. "Why do you tow the 170 onto the step with the boat?" he asked. And then he seemed to telegraph his willingness to help us out. "Is it just to shorten the takeoff run?"

Gordon Mitchell and I both recognized the light at the end of the tunnel. In unison we replied, "Yes! It is to shorten the takeoff." The inspector could easily see the big map on Mitchell's wall, and it was obvious that Schist Lake offered no shortage of takeoff distance; that lake must have provided five miles of open water ahead of any departing aircraft.

The inspector turned to face us and said, "Well, I don't see a problem with safety from your towing operation, so I suppose we can approve it so long as it is just for shortening your takeoff run." Gordon Mitchell and I heaved a heavy sigh of relief. That big clock on the office wall no longer sounded deafening, and my heart stopped pounding.

I don't know when Mitchell became aware of it, but thirty-seven years passed before I learned the truth behind this story. I was enlightened by aviation historian Bruce Gowans. Maybe I should have guessed that because Mercury is an American company, the photograph of CF-HIY and Gordon's speedboat first appeared in the U.S. It had come to the attention of the Federal Aviation Authority (FAA). The FAA immediately set about evaluating the safety aspect of such an operation. They conducted their own experiments and had approved our aircraft towing methods, just as Mitchell had performed it. Transport Canada often rubber-stamps FAA approvals, and therefore our own Department of Transport air regulations inspector was enjoying a bit of fun, at our expense.

A week after this fright, my assignment was to fly a geologist into various northern Manitoba lakes and then wait until he was ready to move to his next location. We were introduced, but I forgot his name almost at once, due to the pressures of getting under way. As the afternoon wore on, the wind began to slacken until it had fallen to zero and the entire lake surface was glassy water.

I had built 400 hours of flying time in total, nearly two-thirds of those hours on floats. Oh, the confidence of youth. I believed that I knew all anyone needed to know about glassy water. After all, I had obtained a seaplane endorsement. It proclaimed to the world that I had mastered all phases of seaplane flying, which included glassy water takeoffs and landings. I was soon to realize that I still had a few lessons to learn.

I taxied Cessna 170 CF-HIY out on the smooth surface, and after a magneto check, I raised the water rudders and applied full takeoff power. Our load was a bit heavier than I had anticipated, and I had some difficulty in getting HIY on the step. After a fairly long run while I attempted to "rock" us onto the step, the cylinder head temperature rose near the red arc on its gauge. I closed the throttle in order to allow temperatures of both oil and cylinder heads to cool down for another attempt.

I had noticed that my passenger kept north on his map directly in front of him, instead of orienting map north to coincide with geographic north. He seemed unfamiliar with my pilot's idea of the proper use of a map; I was surprised then when he asked me if I would mind a suggestion about getting this airplane onto the step. "No," I replied, "Not at all. I think I could use all the help and advice I can get today."

"I notice that you are moving the control column pretty far forward and then back, while trying to rock her onto the step," he said and looked out the window for a moment before continuing. "I suggest you haul the column all the way back and watch your bow wave. When it stops moving, come forward with the stick, but only about one inch forward of neutral, rather than all the way to the instrument panel."

I repeated his suggestion and he proceeded with the lesson. "Y'see, what happens when you push forward to the panel is this." He made a gesture with his hand to emphasize. "The float bottoms smack the water and bounce right off again." He went on to say that moving the control column just slightly forward of neutral would allow the float bottoms to contact the surface of the water more gently and without the bounce imparted by forcing the nose too low. It seemed perfectly reasonable; I decided to give his suggestion a try.

While we were waiting for temperatures to stabilize, he offered more advice. He told me that if we didn't get on the step immediately, I should keep forward pressure on the control column with a gentle rocking motion. "Everything in this world has a natural resonant frequency," he said. Due to my background as a commercial radio operator and all the years of ham radio before becoming a pilot, I believed him. "Keeping that in mind," he went on, "It shouldn't take you long to find the frequency at which any airplane on floats wants to rock. All you need do is help it at the proper time. You will find her walking right up onto the step in short order, especially on glassy water."

I sorted through this information in my mind, and the more I considered it, the more reasonable it seemed.

"Once you have gotten on the step," he continued, "you will be able to find an angle between the float bottoms and the water which will yield the best acceleration rate." I understood his point. "If your nose is too high, the heels of your floats will cause an increase in drag. On the other hand, if your nose is too low, the floats tend to plough through the water and, again, increase drag. When you locate that exact angle, you will feel increased pressure between your back and the seat cushion." He sounded for all the world like the next-door neighbor, Wilson, on the television show *Home Improvement*.

Temperatures were back to normal and we were ready for another takeoff attempt. I raised the water rudders and increased throttle to takeoff power. Control column back; watch the bow wave move back until it can move no further; forward, about an inch beyond neutral. I was rewarded almost at once, with a rise onto the step.

Seeking that elusive angle between float and water for best acceleration, I was able, seconds later, to become airborne. We were climbing out to the east. I turned to my passenger with a comment about how well his suggestion had worked. I asked him where he had picked up the knowledge. "Well," he said, "I used to do a bit of flying myself."

Because I had just more than 400 hours of flying in total, I was always interested in learning the number of hours other pilots had acquired. I asked him how many hours he had. He hesitated as though he'd not heard the question. Then he said, "Well, I stopped entering it when I reached 22,000 hours."

I was in shock. My little 400 hours, against his 22,000? "Here," I said, "You'd better fly this airplane." My passenger declined and said I was doing just fine.

"Many bush pilots, with much more time than you have gained, still don't understand how to get a seaplane onto the step the efficient way," he told me. "Especially when it comes to glassy water. Many of them return to the dock to remove part of their load. As you can see, that is really not necessary."

We made several more stops, and each time his method proved to be better than the last, as I gained experience and built more confidence. I was amazed how much better the aircraft performed when it was allowed to operate in accordance with natural laws.

When we returned to Flin Flon, my passenger boarded a taxi into town. I told Parsons about my experience and asked him if he knew who my passenger was. "Oh, sure, that was Shorty Holden. Shorty has probably forgotten more about flying, especially seaplane flying, than most of us will ever learn." Hank respected him highly. I never had the opportunity of thanking Shorty Holden properly for the lesson he taught me on that day so long ago, but his efforts were not wasted. Up to the day when I made my last takeoff in a seaplane, I continued to polish the technique. I was never again obliged to return to the dock for removal of part of my load, no matter the conditions of wind or water. I feel sure Shorty would be happy to know that I passed this knowledge along to a new crop of pilots over the rest of my years of flying.

* * *

One day, I was introduced to a customer from the United States. I was told his name was Don Clark. Don would need CF-HIY for several days to travel into various areas for geological surveys. We enjoyed each other's company for the following two weeks. Once, we had occasion to stop to look at an area that was like a river delta. Rather than a river, however, this was more like a series of small creeks; they were fairly deep but not very wide, and they meandered like serpents through marshy land east of Flin Flon. While attempting to dock on the bank of one of these waterways, I took my paddle from its port float clips, and pushed against the bank. The sodden ground gave way and I was dunked unceremoniously into about six feet of algae-laden water. I remember Don's hearty laughter at my drowned-rat appearance as he lined up his 35 mm camera for a shot.

At least once a day, Don told me about a book with which he would never part. *Song of the Sky*, written by navigator Guy Murchie, accounted for Don's knowledge of navigation and weather-related subjects. Don imparted various gems of wisdom gleaned from its pages. He spoke with authority on Kansas twisters, navigational tricks of the Pacific Island dwellers and many other subjects. He was the perfect instructor. After completing our flight, I drove him to the Flin Flon airport to catch the TransAir DC-3 flight to Winnipeg. Standing outside the tiny local terminal building before boarding his flight, Don gave me a package, asking that I not open it until his DC-3 was on its way. When he had departed, I opened the package. Inside was a copy of *Song of the Sky* inscribed on the title page:

> To Al Williams, able and congenial skipper and bush pilot, from
> Don Clark,
> Flin Flon,
> Professor of Mining Engineering,
> August 1956
> University of Wisconsin
> Madison, Wis.

To this day, that book is one of my most prized possessions.

5 Archie and Ingy

Not long after Don Clark flew with me, I had a trip into No Rum Lake to pick up Frank Doolan. No Rum Lake was named by Archie Talbot and Ingy Bjorensen when a young pilot had left them stranded there for several days. Archie and Ingy were prospecting partners out of Flin Flon. They spent a great deal of their summers and a fair amount of winter seeking their fortune in gold, zinc, silver and nickel in northern Manitoba and northeastern Saskatchewan.

Archie Talbot and Ingy Bjorensen didn't limit themselves to the provinces only; the territories, too, were in their plans.

September weather had already cooled when Archie and Ingy stepped from their dark green Land Rover and walked into the Parsons Airways Northern Ltd. office, half a mile west of the gravel airstrip that served Canadian Pacific Air Lines and later TransAir's DC-3 service. Archie held a map that highlighted a section of northern Saskatchewan. They wanted to fly into a tiny, nameless lake tucked away north of Pelican Narrows. Folks referred to it locally as Hidden Lake. Hank Parsons was preparing for a trip with another client, and although he and Archie were close friends, he gave the flight to one of his pilots, Bart Aldershot (not his real name). This all happened a couple of years before my time, perhaps in 1953 or 1954.

Bart studied the map and decided to carry only the minimum amount of fuel needed. The lake was small and he could see no advantage in carrying excess fuel that might prevent his takeoff. The flight was uneventful. Bart was happy to discover that right in front of the proposed campsite there was a very large flat rock that sloped down into the water. Just below the waterline, the rock had a steep angle, which made it a perfect place for docking the port side of his left float.

"Bart," said Archie, "you make sure you write a pickup for us in the book for Thursday at 14:00 hours."

"No problem, Archie," he replied, "That's only three days from now. I'll remember even if I don't write it in the book."

"Well, don't forget, 'cause we only brought enough grub to last till Thursday night."

Bart taxied out to the northwest corner of the lake. He turned around and took off past the big flat rock at the east end of the lake. Maybe it was just that he was young and inexperienced, or maybe it was because he had a great deal of charter flying to do. Whatever the reason, Bart forgot to record his pickup date. At noon on Monday he was having lunch in the local cafe when he overheard a conversation. One man said, "I wonder what's happened to Archie and Ingy. I haven't seem them around for these past few days. Any idea where they've gone?"

"By George," said the other, "Now you mention it, Frank, I don't know either, but it could be that they're out chasin' minerals again." Bart felt as though he had been hit in the belly with a baseball bat. He had forgotten all about Archie and Ingy.

He left a half-eaten lunch to race to Parsons' office to tell Hank of his unforgivable sin. He hesitantly told his boss the truth. The last of the prospectors' food would have been eaten Thursday night. The weekend had come and gone while Talbot and Bjorensen waited and wondered what had happened to their pilot. Parsons and Talbot's long friendship prompted Hank to make the flight himself. "I'll go to pick them up—if you go, I'll be shy one pilot. Archie'll kill ya, and I can't afford losing even one pilot right now," he said.

En route to the lake, Parsons pondered the best approach to explain the gross error. With some trepidation, he finally decided to take a lighthearted approach with his old friend Talbot. As his left float came to rest against the big flat rock, Parsons opened the window in the left door and said, "Well, I guess you pot-lickers were starting to feel pretty hungry out here with no food."

"No food? Hell. We had no rum!" complained Talbot. Both prospectors were seasoned bush men and had survived on a varied diet of fish, birds, rabbit and other wildlife. There was of course nothing they could do about the lack of rum. To this day, the lake is called No Rum Lake.

One afternoon we received a radio call from a fisherman who had a camp on the Churchill River some sixty-five miles north of Flin Flon, just a few miles west of Pukatawagan. He had a 300-pound fish for me to pick up and transport to a fish packing plant on Amisk

(Beaver) Lake just west of Flin Flon. I assumed that he had 300 pounds of fish to be picked up. Cessna 170 CF-HIY was the ideal aircraft for the job, because 300 pounds was near its normal capacity and would cost the customer less.

When I arrived at his fishing camp, I was astounded. He did indeed have a 300-pound fish—a sturgeon. I had never seen such a huge fish; it was quite a struggle to load that giant fish onto a tarpaulin, which I had laid out on CF-HIY's floor. Then I was off to Amisk Lake and the fish-packing plant at Denare Beach.

Arriving at the packing plant, I found that it was closed. I had been told that sturgeon rots quickly, so my priority was to find someone to help me get this very costly load on ice. After much running around, I was able to locate the manager of the packing plant and place that sturgeon into the ice house. I arrived back at our Schist Lake base just as the sun was sinking into the western hills. I was later told that the 300-pound sturgeon was not a large one. They apparently can grow to ten feet or more and weigh well over 1,000 pounds.

For quite some time after Hank Parsons flew his new Cessna back to our base, he jealously guarded it. At day's end, he'd remove the keys and lock the doors. During this period, I was flying both Cessna 170 CF-HIY and Stinson SR-9 CF-BGS. Jimmy Hoglander would stand on the dock and wait for me to come alongside with HIY while Ernie Day stood on the other side of the dock, removing the ropes from BGS. It was only necessary for me to step out of one airplane and run to the other.

As soon as I had departed with BGS, Jimmy and Ernie could begin loading HIY for my next trip. When I had completed that trip, they would get to work on loading BGS for my next flight. There were times when I traded airplanes four to six times during the course of a single day.

At first I found this to be disorienting, because forward visibility from BGS was virtually nonexistent while HIY provided good forward visibility, comparable to a car or truck. Access to pilot controls also varied greatly between the two airplanes. HIY used a long handle protruding forward between the seats to actuate the flaps; BGS used a vacuum to activate them; a valve (resembling an electrical toggle switch) was mounted on the panel, for flap control.

The fuel selector for the two wing tanks in the Cessna 170 was mounted on the floor in plain sight just forward of the flap lever. The BGS fuel selector for the four wing tanks was buried in a well (with a lid) on the right side of the right rudder peddle. A pilot had to reach blindly into that well and turn a stiff knob to switch tanks.

In those bushplanes it was necessary to change the fuel selector quite often, in order to balance the fuel load. A pilot first selected either starboard outboard tank or port outboard tank and then alternated the fuel supply from those tanks until both were virtually empty. Otherwise, the airplane needs constant left or right aileron to maintain level wings, as there was no aileron trim control.

Our Stinson SR-9 was a member of the Gullwing Stinson family, which originated in the Cord (automobile) factory. Cord also owned the Lycoming Engine Company and often fitted their aircraft with Lycoming engines. Alternate power plants were available as a higher priced option. CF-BGS was equipped with a Pratt & Whitney R-985 "wasp junior" engine rated at 450 horsepower. Another engine option was the Wright R-760E-2, which produced 350 horsepower and was lighter than the Pratt & Whitney engine. The SR series (Stinson Reliant) had its origin in the Detroiter Junior from 1928. The Reliant first sprouted a gull wing in 1936. A military navigation trainer, designated V-77 was produced during World War II. The SR family shared a structure of welded steel tubing including the Monospar wing. The wing ribs, however, were of aluminum. All the aircraft had a fabric covering stretched over suitable wooden fairings.

My first sighting of a Gullwing Stinson was at Prince Albert on the North Saskatchewan River in 1938, when I was nine years old. Little did I know that some eighteen years later, I would captain her sister ship. She was an SR-8B, registration CF-BGW. A Lycoming R-680 engine of about 300 horsepower was used in that airplane. CF-BGW had been wrecked in the bush by Tom Lamb at Cumberland House due to engine failure. M&C Aviation, of Prince Albert, purchased the remains and rebuilt her. On June 28, 1944, old Stinson SR-8B CF-BGW crashed at Emma Lake in northern Saskatchewan and was destroyed.

After the novelty of owning a new aircraft wore off for Hank Parsons, I was finally allowed to fly his Cessna 180. First I flew CF-HDX and then

IXU; these aircraft as well as another 180 were fitted with Sunair S-5-DTR radio transceivers. Hank Parsons was of the old school; he placed very little value on modem technology. To Hank, the pilot who needed a radio in an airplane was not really much of a pilot. "I don't want you pot-lickers using them radios all the time; you'll wear 'em out," he told me.

"Hank, if I can't use the radio, just take it out, I can use that extra weight for my customers," I responded. We continued to use the radios.

Manitoba Telephone System maintained a radio circuit out of The Pas, and virtually everybody monitored that frequency. It was a natural for pilots to periodically provide position reports. MTS would acknowledge the position report. Should anything go wrong, they would know the last position of that aircraft. There was another reason for giving these reports: any would-be customer, in any of the various camps, could "flag down" an airplane that happened to be in his vicinity.

I once had a charter to Snow Lake, and as I departed the lake at Snow Lake, I reported the fact on MTS radio. Almost at once a call was received from a camp about seven miles south of my course for a charter to Flin Flon.

The Air Transport Board, our government's watchdog over fair air charter competition, dictated that a charter would be charged from its point of origination to its destination and return. If a pilot arrived at a destination and new passengers wanted to charter, a second, round-trip charter fee was applied to the new passenger(s).

Because of the ATB ruling, there was no alternative to me charging a second trip from Snow Lake as a round trip from Flin Flon to that camp and return to Flin Flon. To do otherwise would have been a contravention of Canadian law.

I reported the results of utilizing the radio to gain double charter income from the Snow Lake flight to Hank Parsons. He immediately called the pilots together to establish double charters through radio as a new company policy. Bert Warttig, the pilot whose crash had resulted in me being hired by Parsons, was now back with us after about a seven-month absence. Jimmy Hoglander had graduated from helper to commercial pilot (he had been building time in Parsons' de Havilland Foxmoth).

Parsons was never a man to pass up an idea that would generate additional revenue. This was understandable considering the fact that sometimes a charter could not be completed because of weather or mechanical problems. The occasional double charter would more than offset our nonrevenue flights.

During my thousands of hours as a commercial pilot, I was most fortunate to have flown thirty-four aircraft types; this list includes several Beeches, Piper and many models of Cessna aircraft. I also flew several de Havillands, including Standard Otter, Tiger Moth and Beaver. I also piloted an Aeronca Champion, Chief, Douglas DC-3 and Noorduyn Mark VI Norseman, as well as a Luscombe Silvaire. I even flew a Helio Courier.

With ratings of "Single and Multi-Engine Land and Sea" it was natural that 1 should specialize in bush and Arctic flying. I eventually added a ski endorsement for winter operations and a night endorsement. It wasn't long before I realized I still preferred hot food to cold food whenever I was far away from someone's kitchen, or was on a long bush or tundra charter. While canned peaches provided me with needed liquid, I wondered if I could warm up canned goods such as beans by placing it on the hot engine just after shut down. Unfortunately, after air-cooled engines stop, they don't retain heat long enough to cook anything placed on or near them.

I pondered the problem for several days. Because of my years as a ham radio operator (radio men are notorious for experimenting), I continued to refine my original idea of cooking on a hot engine. I decided to place a can of beans between two cylinders while the aircraft was in flight. However, after placing a can of beans between engine cylinders, I was directed to fly to a medical emergency north of Wabowden. For the next two hours during the emergency flight, I couldn't help wondering what was happening to the beans under the engine cowling. I expected to hear, any second, a mild explosion accompanied by the odor of burned beans. After landing and shutting down, while my medical passenger was being settled comfortably, I carefully opened the engine access door outside the airplane. I discovered an intact air-cooled can of beans, its temperature perfect for eating. Subsequent, shorter flights proved that beans, or any other canned food, could be heated in just a few minutes but would never overheat when placed on the engine. This was only true during the summer.

During that same summer, Hank Parsons decided to reactivate his scheduled charter flights out of Flin Flon to Snow Lake. He used his pride and joy, the newly purchased CF-IXU. People trusted Hank Parsons. They began to depend on his "sked" for routine Tuesday and Thursday flights. It wasn't long before his sked flights were bringing in revenue at much higher rates than regular charters. At twenty cents per passenger mile and full seats, a higher rate was earned than the normal forty-five cents per mile charter. Added to the sked revenue was a mail contract and ten cents a pound for express shipments with a minimum charge of one dollar per package.

With a Snow Lake base, Parsons Airways Northern could charter from the base and save the customer a return ferry flight charge from Flin Flon. A charter simply waited until Tuesday or Thursday to realize great savings, due to air transport regulations, on their air transportation costs; otherwise they would have to charter from Flin Flon.

Before I flew Cessna 180 CF-HDX on a sked run, I had never had a chance to fly a canoe tied to the side of float struts. Cessna 170 CF-HIY simply didn't have the necessary power to fly anything tied outboard. Gullwing Stinson CF-BGS had not been used, as far as I knew, for the purpose. Although I later discovered Hank had flown a few on her.

CF-HDX, like IXU, was equipped with canoe ropes, although I hadn't used them until that particular day when Parsons asked me to take the sked.

Just before I left the dock, Parsons asked if I had canoe ropes with me. I told him I did.

I arrived at Snow Lake to unload passengers, baggage, mail and express as our agent told me he had a charter for me into Reed Lake. The clients asked if I could handle a canoe for them. Naturally I said I could fly their canoe. After all, I was set with canoe ropes. I also knew that square-stern canoes, like the one they had, were always flown stern first. Hank Parsons had told me canoes were flown stern first due to convention, not aerodynamics. Always the experimenter, I decided to question Hank's theory on another day. Air piles onto the square stern and spills around the edges, creating a near laminar flow, along the smooth canoe sides. In this way, drag could be reduced, I reasoned.

Tying the canoe proved to be more difficult than I remembered from watching Hank do the same job. Of course, I had not paid close attention. Due to the beam width of the canoe, I found that I could not open the left cabin door with the canoe tied in place, so I turned HDX around at the dock and entered through the right door. The clients wanted to fly out on the same trip as the canoe, but I said that they would have to wait, along with their gear, for a second flight.

As I taxied out on Snow Lake and lined up, into wind, parallel with the shoreline, HDX handled as though there was no canoe tied to her floats. During the normal level attitude of taxiing, the canoe was above the waterline. When I opened the throttle for takeoff and hauled back on the control column, however, the nose came up and the float heels were lowered below the average level of water. Due to the canoe's overhang behind the float heels, the tip of the canoe trailed in the water to act as a rudder. HDX slewed to the left. I was suddenly 90 degrees out of the wind, facing the eastern shoreline; I was forced to chop power. I realized what was happening, but I thought it was probably normal under the circumstances.

I taxied out to try again, but his time I began my takeoff run cross wind, away from the shore. By the time I had HDX on the step I had come around parallel to the shore but was airborne in fairly short order. I felt a bit apprehensive in the air, because periodically HDX would jump sideways. I knew the cause: wind gusts were catching the canoe, driving us sideways. I didn't realize this was anything beyond what one would normally expect. Ignorance is bliss.

When I landed at Reed Lake, I experienced that same ninety-degree veer to the left as HDX came off the step. Again, I didn't have a clue that anything was wrong. I untied the canoe and put it on shore, tying it to a tree in case the wind should blow it free. Then I returned for the passengers.

When I left the men and their gear on Reed Lake, I was asked to write them in for pick-up on Thursday's sked. Parsons had just arranged a short-term lease of a de Havilland Beaver, and on Wednesday he told me he would fly Thursday's sked in that aircraft. I told him to be sure to pick up the men at Reed on Thursday. I was away when Hank returned from his sked but next morning he called me into the office. "Did you fly a canoe into Reed Lake a couple of days ago?" I replied that I had. "Well," he said, "Them pot-lickers know we can't fly a freighter canoe on a Cessna 180. How in hell did you get that damn thing airborne?" Hank paused for a moment. "1 had to leave it there. It's too big even for a Beaver, and I've hauled eighteen-footers on a Beaver; we have to wait for a Norseman to go in to get it out."

The old adage "nothing worth learning can be taught" certainly applied to this situation.

A radio call from a geophysical group camped about forty-two miles south of Flin Flon with a request for transportation supplied my next adventure. They had rigged up an A-frame from which they had intended to drive a four-inch pipe into the ground for use in their work. They had ordered a cast-iron pile driver weight from a Winnipeg foundry. This device was about two feet in length and several inches in diameter and weighed in at 380 pounds. The weight was painted bright blue, with the section near the top in red. This would make it easier for the crew to see it when in use and avoid physical contact with it as it fell from the A-frame to drive the pipe into the ground.

The camp we delivered the weight to was approximately three-quarters of a mile from the lake. The question was, how were they going to transport such a heavy object from the lakeshore to their camp?

I had recalled reading about how missionary pilot Nate Saint, flying in the jungles of Ecuador, would deliver medicines and other supplies to remote jungle clearings. As he lowered a canvas bucket at the end of about 1,500 feet of line, Nate would maintain a fairly tight turn to remain over the selected clearing. The aircraft, line and bucket would form a cone, with the aircraft orbiting the rim of the cone, the line forming the narrowing sides and the canvas bucket stationary at the apex of the inverted cone. As the bucket dropped, it would lose all horizontal motion to come to rest in the middle of an open space in the jungle.

It seemed to me that if Saint could accomplish that with a Stinson 108-2 Voyager, I could do something similar with a Cessna 180.

The pile driver weight was delivered to our dock, and I radioed the crew. Our discussion settled the matter of delivery. We would remove the right-hand door from Cessna 180 CF-HDX and Ernie Day would secure himself behind the right seat, which had also been removed. Ernie made a rope harness and secured it to seatbelt anchors so that he could throw the weight out the door at the appropriate time and maintain his own safety.

The crew told me there was a clearing directly in front of their camp; they would remain well to the camp side of the clearing. If possible, I would drop the pile driver somewhere near the center of the clearing. With Ernie secured in his web of rope, we took off with the understanding that I would lower 40 degrees of flap and, as we came over the clearing, I would do a very steep turn (about 85 degrees of bank) to the right. Ernie would then push the red and blue weight through the right cabin doorway. If our calculations were correct, our cargo should land very near the middle of the clearing.

As we approached at fifty feet with 40 degrees of flap over the center of the open area, 1 placed HDX in a near-vertical bank and Ernie forced the heavy weight through the door. Shortly after making the drop, I asked the crew by radio how it had gone. "HDX, Camp Three," I was told, "A direct hit on the center of the clearing ... thank you. "

Ernie and I returned to Schist Lake brimming with pride that we had accomplished our task. Later in the day, I had a trip to a prospector's camp. While I was airborne, the geophysical camp called me on the Manitoba Telephone System Radio Frequency. "HDX, Camp Three," he said, "You did a good job of hitting your target—but we didn't realize that the ground was nearly as soft as it turned out to be. Ahh ... we're still probing for our weight, but so far we haven't been able to go that deep. We think that weight may still be going down."

Thus ended our one and only attempt to airdrop such heavy freight.

A few days later, I was doing the sked run when I stopped at the Snow Lake dock, where Vic Olson was waiting for me to arrive. He had a charter for me. Vic Olson and I had known each other for over a year. Vic was the purchasing agent for Britannia Mines at Snow Lake. He had been a Greyhound bus driver where he'd worked with my friend Roy Sanderson and my brother-in-law "Cardy" Rollefson. He approached me at the dock with a load of groceries that he wanted flown out to a camp at Tramping Lake. Nelson Hogg and his crew were exploring mineral deposits that might be suitable for mining. Britannia Mining was a subsidiary of Howe Sound Mining & Smelting, whose Britannia Beach operations were not far north of Vancouver.

Nelson Hogg was a good man, without question. Nelson would never place anyone's life at risk intentionally. He was simply a man trying to do his best for his company, but he had little experience with bush life.

Hogg was the ultimate company man, always wanting to negotiate the best deal he could for his company. An admirable trait unless you happen to be on the receiving end of negotiations.

Our pilots, Bert Warttig, Jimmy Hoglander, Rudy Hoffman and Ian Adams, were getting tired of arguing about loads they were willing to carry. Their Cessna 180s were equipped with Edo 2870 floats. (This number is indicative of floatation ability.) Nelson always wanted pilots to take a larger load. He was also a man who didn't like flying very much, and I'm sure he waged an inner battle, trying to secure as big a load for his company as possible while being leery about his own safety.

On this occasion he needed a Cessna 180 to haul a load out of Amisk Lake (Beaver Lake), a few miles west of Flin Flon. I was sent to fly the load in HDX. This Cessna was equipped with the much smaller Edo 2425 floats—a fact that Nelson Hogg overlooked. I decided to try to cure him of constantly hounding us for bigger loads. We began loading, and as the aircraft took on more load, the floats began to sink lower in the water. Before we had what I considered a full load, Nelson said, "That's pretty well a load, don't you think?"

"Oh, I guess we can take more than that. Hand me that box over there and I can place that bag just behind the cabin door and the duffle bag by your foot—we can put that just behind the seats." Nelson looked rather uncomfortable. This was his neck at risk.

Finally I said, "Okay Nelson, I'd say that's a load."

Hogg stepped onto the starboard float, and its deck immediately became awash; he jumped back onto the dock. "We'd better take some of the load back off," he suggested. "No. That's okay, Nelson. The float will only go under from the rear spreader back. It'll pop right back up once you get in." As we taxied out onto the lake, Nelson asked several times how the aircraft handled with its big load (actually, we had no more load than was normal for these small floats).

Our takeoff was done with lower power settings than normal, and the takeoff run was somewhat longer than Hogg expected. He tapped me on the shoulder. "How's it going? We too heavy?"

"Oh, it may be bit heavy," I told him, "But there's no problem. I always carry a full load for my customers." Nelson Hogg never again argued with me or any other pilot, that I'm aware of, about putting on more load.

Vic Olson and I dropped our groceries at Tramping Lake and then returned to Snow Lake. Nelson and his crew went into the camp and, in addition to personal belongings, took in their fresh meat supply. The crew was a good group of experienced bush men. They were not, however, happy campers.

Along with their fresh meat, Hogg had insisted on taking along a roll of moldy bologna which the men had refused to eat. That bologna came from their previous campsite. It should have been obvious that the crew was not going to eat it at their new location. To add insult to injury, Nelson's limited experience of bush camps resulted in the meat supply being stashed atop a tree limb about ten feet from the ground. No amount of arguing from his crew would convince him that a bear could easily climb the tree. The men wanted to suspend the meat from a limb so that the bear could neither reach it from ground nor from tree. As Nelson was in charge, the pack was deposited on the tree limb. During the night, a bear visited the sleeping camp and consumed the fresh meat. It left the moldy bologna for the crew. An order of fresh meat was flown in later that day and was suspended from a limb so it could not be reached from above or below by any bear. The moldy bologna was finally thrown out.

Flying out of Flin Flon each day provided us with a wide range of territory over which to roam. We usually stayed north of Suggi Lake in the south but flew to Brochet (on the extreme north end of Reindeer Lake) and beyond to the southern reaches of the Northwest Territories.

In the morning I could be at Lac La Ronge, where I sometimes visited with my old friend and mentor Tom McRae. He was a radio amateur, VE5TM, and got me started in that hobby. Tom was employed as a point-to-point radio telegraph operator for the Saskatchewan Department of Natural Resources at La Ronge. That same afternoon I might be at Island Lake, Oxford House or God's Lake, near the Manitoba-Ontario border. Flying at 130 miles per hour, we could and often did touch down at spots hundreds of miles from our home base on Schist Lake. More often than not, we managed to make it home at night, like a flock of homing pigeons.

Many times during that summer Cornelius Vanderbuilt Whitney flew with me as a passenger on a variety of trips across northern Manitoba. Mr. Whitney was owner of Hudson Bay Mining & Smelting's exploration sites. He also owned a Hollywood film studio that produced movies in which his wife often had a starring role.

Hudson Bay Mining and Smelting based a fleet of aircraft at Flin Flon that included a de Havilland Otter, a Norseman and a Sikorsky S-55 helicopter. They would normally be engaged in transporting material and staff to many job sites throughout the area. The company collective agreement on air service called for a company aircraft and a pilot to stand by for emergency flying on the weekends or holidays.

Cornelius Vanderbilt Whitney would often arrive at Flin Flon's landing strip on a weekend in his personal transport aircraft with a desire to visit some of the mining areas that were being developed. His air service agreement made it necessary that he call for a commercial aircraft, and it just happened that I was often the pilot with whom he flew.

We flew together enough that Whitney felt very comfortable with me at the controls. Before many trips had been done, we ended up on a first-name basis. He called me Al and insisted that I call him Corney.

I cannot recall a time when we didn't have an aircraft available for his use. After a few trips, he began to request that I pilot him to his destinations.

Whitney could be a generous man. A short time prior to my arrival in Flin Flon a fire had raced through the community center, totally destroying the building. The citizens of Flin Flon began fundraising efforts to replace the center. A few days before Christmas, during the opening ceremonies, Whitney stepped onto the stage. He announced that he recognized the marvelous work which the citizens of Flin Flon had done in fundraising, though he knew that a large sum was still outstanding. He then presented a check for $2 million to the mayor to cover the remaining debt!

Because Flin Flon is close to parallel 54, the larger lakes are very slow to warm during summer. Local kids had no accessible swimming hole until Whitney sent his company bulldozers to open trails (during winter, when the ground was frozen) to a small lake close to town. Phantom Lake, as it was called, was sun-warmed early in the season. Its weed-infested and muddy bottom was also cleaned up with truckloads of gravel and sand, laced with a mild weed killer, which was spread on the winter ice surface to sink to the bottom in spring.

During my time at Flin Flon, the excellent labor relations between the company and its employees amazed me. I was told that union meetings were more likely to discuss what activities should be arranged for the First of July celebrations than working on demands to present to the company. Employee turnover was minimal. Flin Flon had the highest per-capita wage structure in the whole of Canada at that time—in spite of very stiff competition from other producers.

During one of our flights I mentioned the high wages, community center donation and the Phantom Lake project. "While I really did want to see kids with a nice place to swim, it is still good, sound business as well," Corney told me. He knew that a happy work force made for production and cooperation. Present-day management could benefit from his wisdom.

* * *

I enjoyed flying Cessna 180 CF-HDX. She was one of the earliest models of 180. She was equipped with a Continental engine of slightly less power than that of IXU. Whereas IXU developed 230 horsepower for takeoff, HDX delivered only 225. IXU and all newer Cessna 180s I flew were equipped with the McCauley constant speed propeller, while HDX used a Hartzell constant speed prop; I believe both manufacturers made the props seventy-four inches, but Hartzells were designed differently.

Hartzell propellers wanted to go into fine pitch from twisting moments exerted on the blades by the airloads. Oil pressure from the engine caused the propeller governor to force the blades into coarse pitch during cruise.

My personal experience backs up this statement. One day I took off from a lake near Pelican Narrows. Just after takeoff, I lost oil pressure to the propeller governor. The prop went into fine pitch and I was able to limp back to Flin Flon. Flying with the propeller in full fine pitch is quite similar to driving your car along the highway in low gear.

The McCauley seemed to work in reverse when compared with the Hartzell. It would go into coarse pitch if oil pressure was lost. This is like trying to climb a hill with a car in high gear— you may not make it home and you can cause a lot of engine damage, perhaps blowing cylinder heads or caving in pistons.

Many of the older Cessna 180s had extra bracing wires installed diagonally across the windshield that crossed each other in the center of the area above the instrument panel. HDX in all likelihood had these wires added after I flew her; when I flew her, the windshield would work its way vertically and horizontally within its frame. In addition, the engine mounts were not as well isolated as they were in later models, so that the pilot had more of a feel of engine vibration through the control column and rudder peddles.

This gave me the impression that I was in much closer contact with the power-plant in HDX than in other airplanes that I flew. Our Gullwing Stinson SR-9, BGS, was driven by a Pratt & Whitney R-985, 450-horsepower engine. Her flaps were operated by vacuum, which either fully retracted or fully extended them. After a time I discovered that those flaps could be made to extend to midway between the two extremes by retracting them after takeoff and allowing forward speed to build to around 100 miles per hour, then selecting full flap again. The system could only exert sufficient force to droop them about half way.

With only about half flap, BGS was a good climber, but it took me some time to discover this fact. In the meantime she used to scare me half to death during some of those takeoffs, when the trees grew taller and taller as I approached the shoreline. BGS was also the only airplane I had ever seen with a control wheel that was perfectly round. It was constructed of fine wood. Its fit and finish would have made Rolls Royce engineers proud.

I once used CF-BGS to pick up some American sports fishermen from the camp they had set up near Hanson Lake. They must have arrived at our seaplane base directly from Channing Airport, without having been to town. When I deposited them back at Parsons dock on Schist Lake and extended the chance for a ride to town, they agreed. As we crossed Ross Lake on the road into town, we topped the hill where Flin Flon came into view and one of them turned to the others. "Hey fellas," he said, "We're in Fling Flong!"

Until that moment, I had never given any thought to Flin Flon's name. Hank Parsons told me this story: Tom Creighton, a northern prospector, happened on an old abandoned campsite not far from the Churchill River, in 1913. Creighton and his companions found and began to read a tattered science-fiction novel called *The Sunless City*.

A character named Josiah Flintabbatey Flonatin finds a bottomless lake, and in order to prove that it really is bottomless, he constructs a submarine to find himself in the centre of the earth. Eventually he escapes a matriarchal society that controls a pure gold city. The last three novel pages were missing, so Tom's crew never discovered how the tale ended.

A short time later that winter, Tom ventured out onto Ross Lake and the ice broke under his weight. Soaking wet, he managed to scramble to shore and started a fire in a rock crevice to keep warm while his clothes dried. When the fire melted the ice and snow, it was discovered that the crevice was loaded with minerals. The men began staking claims around this find; they decided to name the area after Josiah Flintabbatey Flonatin. The name, of course, was much too long, so someone shortened it to Flin Flon.

6 Winter Flying

At the end of our summer flying, north winds were becoming colder with each day; water froze daily on our rudder cables. We were having a hard time maintaining directional control on the water because the cables would not freely pass though their pulleys.

Pilots climbed out on the floats to free the cables of ice before takeoff. This helped airplanes in flight because water rudder cables were connected to the rudder pedals through coil springs. While an increase in force might be necessary to move those pedals, aircraft safety in the air was not jeopardized.

Summer became fall when lakes froze. Fall, in turn, gave way to winter operations with skis, engine covers and blow pots for heating our engines on frosty mornings. I hadn't done very much winter flying to this point; in fact my only ski experience had been while I was learning to fly with the Winnipeg Flying Club in Winnipeg.

The Winnipeg Flying Club owned a Cessna 170 CF-FBG. My first check ride in FBG was on wheels in November 1954. Len Lavoie had been my instructor. Len and Leo Hoffman were the main instructors at that time, while Herb Taylor was the club manager. I had put in very little time on FBG until February 1955, when I was told that I had to do a cross-country flight.

On February 18, 1955, I took my fellow student Mel Stitt on a local flight in CF-FBG on skis; it was decided that my commercial cross-country trip would take place on the following day, weather permitting.

My wife, Louise, Peggy Elliot, and her mom Gert would be my passengers, with Sioux Lookout, Ontario, our destination. Peg and Gert were both radio amateurs.

I was told the mixture control in the Cessna 170 was ineffective and was advised not to use it. The mixture control in an airplane operates on the same principle as a car's choke, but the application is reversed. The choke in an automobile engine causes a richer mixture of fuel and air so the engine can operate when it is cold. In an aircraft, the control is used to lean the mixture at altitude. Most cars always operate at the same altitude. Airplanes, on the other hand, are continuously changing their height above sea level; the fuel-to-air ratio is in constant need of adjustment.

At the time of our cross-country flight to Sioux Lookout, I had not learned the finer points of fuel management or the science of fuel and air mixture, nor the effect of altitude. I took the advice not to bother with the Cessna's mixture control—and I was to regret it.

February 19 dawned sunny and cold—an ideal day for a trip to the bush country of Northwestern Ontario. As we skirted Beausejour and Whitemouth, Manitoba, and Minaki in Ontario, the outside temperature began to rise as the sky became overcast. By the time we'd reached Niddrie on the CN Rail line, outside temperatures had risen further. Heavy wet snow began to fall and it began to freeze to the wings of CF-FBG.

I wanted to turn back but decided there wasn't enough fuel remaining. As we flew eastward, visibility had slowly become less and I would have had to fly many miles to the west before finding good visibility again. Our fuel consumption had been higher than I had anticipated and thus I felt we might not have sufficient fuel left to fly all the way back to where visibility would allow me to find a good landing spot, and because of the relative cold, I was worried about our survival while awaiting rescue. It seemed that our best chance was to travel the fairly short distance to Sioux Lookout if we could see well enough for that last bit of flying. I was afraid of landing in the soft snow that lay at our feet, because FBG, with its 145-horsepower engine, would never get airborne again. We pressed on, hoping for an improvement. This is a major factor in aircraft fatalities—continued flight into severe weather conditions. If I had leaned the mixture, I might have had sufficient fuel to return, at least as far as Kenora.

We had railroad tracks beneath us, and I was loath to lose sight of them in case we might become hopelessly lost in the near-whiteout conditions. At this point we were no more than 100 feet above those railroad tracks. We quickly passed over Hudson, where the Elliott brothers built their famous wooden aircraft skis often called boards (although I understand the ski pedestals, which attached the skis to landing gear legs, were usually made by Mason and Campbell Aviation [M&C] at Prince Albert). Only about ten miles to go. Almost at once we were over Abram

Lake and the Department of Lands and Forests Air Service Base. I lined up for final approach. This very nearly did become our final approach. I misjudged my altitude and speed, probably the result of being under strain for such a long period. I realized we were not going to touch down before running out of lake.

I rammed the throttle to full power, but FBG did not want to climb. This was the result of ice on the wings, which interfered with the airfoil design. When it seemed inevitable that we would crash into a stand of pine trees on the shoreline, I tried a desperate gamble. I lowered the nose, although we were only fifty feet above the surface, and rolled into a right bank. The gamble paid off, and we narrowly avoided disaster.

When I realized there was ice on the wings, I flew well out toward the far shore and made a gentle turn and approached again. This time I allowed plenty of room and sank slowly until our skis touched the snow. Once we were down, Harry Speight, who was an Air Service pilot, showed me how to ensure that my skis would not freeze to the snow overnight. We walked into the woods and cut down a few spruce boughs. Then I taxied forward to stop with my skis on top of the thin boughs of spruce. Those boughs acted as insulation, isolating my skis from the surface.

Louise and Peggy had to get back to Winnipeg. There was no relief in sight from the snowstorm, so they boarded the train. I think they were glad to avoid any further anxiety from flying over the rugged bush country of northwestern Ontario.

After a three-day storm, Gert Elliot and I were able to leave on a day the sun shone brightly. CF-FBG didn't move without help through the heavy snow pack. Lands and Forests Air Service personnel were able to keep us moving, but when they stopped pushing on our wing struts, FBG came to a stop. Harry Speight was leaving for a trip in de Havilland Otter CF-ODK; he suggested that we might be able to get airborne by taking off in his tracks. Harry applied power, and the 600-horsepower Pratt &Whitney engine moved CF-ODK out onto the lake. The Air Service men pushed on our wing struts, and soon we were in the Otter's deep snow tracks. Once in the Otter's tracks, FBG could move under her own power.

When the snow settled and I could see again, I opened up to full takeoff power, and FBG began to move. We had gained about forty-five miles per hour when Harry's tracks came to an end—the Otter was able to lift off in a much shorter space than was needed by FBG. We found that FBG would bounce into the air but we had insufficient speed. She would settle back onto the surface until the next drift would pole-vault us back into the air. After six bounces, FBG finally decided to remain airborne. Gert and I were relieved to be on our way home. We stopped at Kenora for fuel because we had left Sioux with a reduced load of 80/87 aviation gasoline because of the heavy, soft snow covering. The runway at Kenora had a high spot right at its middle, and I didn't touch down until we were very near that spot. As we came over a well-packed surface in the middle of the runway, I began to doubt the FBG would come to a stop. However, we were able to stop with plenty of runway remaining. We were soon back in Winnipeg.

Of course, that trip had taken place long ago, and now we were to begin winter flying all over again. I had learned many things in the intervening years, but still, winter flying was very different. I was bound to learn more.

During my first year of commercial winter flying with Hank Parsons' air service out of Schist Lake at Channing, Manitoba, I learned a great deal more.

On Christmas morning 1956, around 07:00 hours, I was awakened by the telephone. Hanks' voice was strained as he said, "Bert Aspenol is dying. Nursing station at Moak Lake says he needs to get to a hospital as fast as possible." He paused and then said, "I had a call from Lamb Airways. Lamb tried to get Bert out of Moak Lake, but bad weather turned them back. They called here to see if we can do anything to help. They got as far as Cormorant Lake, but it was in the trees over there."

"I can't go," he said, "because I've been drinking; Bert Warttig is in The Pas and can't get back. That same weather has him trapped." Hank took another breath and went on. "I called Jimmy Hoglander, but he's been drinking, and he's still drinking." One more pause. Then Parsons placed Aspenol's life in my hands. "You're the only one who can get him. Can you make the flight?"

He always asked. Hank would never order a pilot take any flight; the pilot had to be willing to go. Hank had only ordered a pilot to fly once, and the pilot had refused. Parsons had made the ill-advised trip, which nearly ended in disaster. He never again ordered a pilot to fly. A willing pilot resulted in a positive attitude; none of us ever turned down his requests.

I shook my head in an attempt to wake up. "Okay, Hank, what's our weather?" Hank had asked the TransAir agent to send up a weather balloon. It had disappeared around 200 feet. This was not good news, because such a ceiling would make a 200-mile mercy flight a difficult one. I'd have to go. If Aspenol died after I refused to fly, I wouldn't ever be able to face myself in the mirror again. "I'll be there in twenty minutes," I told Hank.

I'd first met Bert Aspenol the previous summer; he was six feet, eight inches tall and 275 pounds. Bert's partner was a small Cockney named Alf. They had boarded Cessna 170 CF-HIY to be flown from Flin Flon to Moak Lake, north of the Burntwood River. They were both drunk. An alcoholic beverage in a Coke bottle was rapidly changing hands. When I noticed this, I asked if the bottle was empty; they assured me that it was. "All right, fellas," I said, "Now that Coke bottle is empty. Have you any more liquor with you?"

"No, that was all we had." They both laughed heartily as though over a private joke.

We taxied out on Schist Lake; I was about to open the throttle for takeoff but first glanced back at my passengers. I spotted a new bottle of Canadian Club whisky passing from Alf's hand into Bert's. My reaction was instantaneous; I snatched the bottle from Bert's hand and popped open the window. Then I emptied the bottle into Schist Lake. I had overreacted, of course, but the sight of that bottle had made me furious. To add insult to injury, I dropped the water rudders, which had been retracted for our takeoff, turned around and taxied directly back to our dock, where I ordered the two drunken men out of HIY as I said, "You fellas come back tomorrow when you're sober. I'll fly you to Moak Lake then."

Bert Aspenol was powerfully built, and for a moment it seemed he might attack me. At six feet, two inches and 225 pounds, I was fairly confident I could handle him, but instead he decided to take the matter up with Hank Parsons. I had only been with Parsons for about two weeks and wasn't too sure about Hank's position on my decision. I needn't have worried, because he stood behind my decision and added, "Now, you two pot-lickers get uptown and come back here at 09:00 tomorrow. You'll be sober, or I'll make sure you don't go then either!"

Next morning, the two arrived dockside, sober and glum, refusing to ride with me. So Parsons was stuck with them. I was somehow disappointed because I have always felt the customer is right unless that customer is obviously wrong—and certainly, showing up drunk for a flight in what promised to be rough air is obviously wrong.

As it turned out, however, they arrived back at their jobs at Moak Lake without any real harm being done. I didn't know it at the time, but Aspenol was the camp cook for Midwest Diamond Drilling, which was test drilling on the site. His meals were of the highest caliber, and I wondered how such a great cook could have found his way into such an isolated drill camp. I was to discover, a few days later, that Bert Aspenol had worked as a chef at the Waldorf Astoria in New York; he could not keep his job, however, whenever he was near booze. In northern Manitoba however, Wabowden, the nearest alcohol outlet, was about thirty minutes away by air.

Moak Lake had been staked by a pair of prospectors who had been poking around for years, and International Nickel Mining Company was doing an extensive search of the area. Joe Kerr, Gordon Crosby and Walter Johnson were the partners involved in prospecting that area.

On this Christmas morning, after heating the engine of Cessna 180 CF-IXU, I taxied onto the ice of Schist Lake and took off for Moak Lake. That reported a ceiling of 200 feet, forty-five minutes earlier, still held as I turned northeast, heading zero seven five degrees, which would carry me to Moak Lake. I discovered soon enough, however, that ice would build on my wings unless I stayed much lower. Before long I had to stay about twenty feet above the trees. I'd climb up one hillside and descend the other side, but those brief excursions into cloud base accumulated ice, so it was necessary to remain near twenty feet above the trees for several minutes to wait for the ice to melt, to regain proper airfoil efficiency.

Navigation at twenty feet above tree tops is difficult, but one hour and forty-five minutes later I arrived at Moak Lake. Bert was brought to IXU. He looked terrible. Bert had jaundice and he had to get to a hospital as quickly as possible; without independent medical knowledge, I had to take the word of nursing station staff on his condition. My job was to get Bert safely back to the Flin Flon Hospital just as soon as I could.

The return trip was similar to the outbound trip; the ceilings stayed low and I experienced the same difficulty with navigation. Being so near the tree tops, my visibility was reduced in snow flurries. This was not the kind of flying I enjoyed.

Manistikwan Lake lay parallel to Schist Lake, about two miles to the east. I had already been advised that in good weather I should time my flight from the center of one of the lakes to the center of the other, because of the hill between those two lakes. On a bad weather flight, such as today, the timed flight knowledge might prevent an accident.

As we reached the parallel lake, the hill was totally obscured. There was no way to fly around it. I dared not land on the lake; we might have to wait for hours, perhaps a day, for a weather improvement. I knew that such a delay would likely finish Bert.

Striking the top of that hill at two miles per minute would almost certainly terminate both of us, but I had to make the effort. I flew back to the east, turned and set up a westerly climb at 120 miles per hour. I checked the second hand on my watch. I would time my flight so we would, I hoped, descend on the other side of the hill to break out of cloud directly over Schist Lake. My timing and speed were accurate, or perhaps a guardian angel rode with us that day.

Five minutes after landing at Schist Lake, I drove Bert Aspenol to the Flin Flon Hospital, where he recovered from his illness and was soon back at Moak Lake turning out those great meals. From that day onward, whenever I arrived for lunch I found I was being treated like visiting royalty. Bert always provided the men with a choice of beef, pork and turkey, prepared to perfection; he always offered several fine desserts. I was almost embarrassed by Bert's attention. It was a far cry from the treatment I had received from him after I had dumped his whisky in Schist Lake during our first meeting.

I made many flights to Moak Lake, Cook Lake and Thicket Portage during that winter. Late one afternoon in Thicket Portage, Roddie McIsaac, who owned Mid-West Diamond Drilling, and two of his men needed a flight back to Flin Flon. Outside temperatures were in the minus-thirty range; darkness was settling in on the surrounding countryside as we came over Setting Lake, just west of Wabowden. McIsaac suddenly realized the truth of the statement, "You can't own beer, you can only rent it." I handed him a waxed-paper cup, which we used as an airsickness container. When he had used the container he wanted to open the starboard window.

"Don't do it, Roddie," I told him. "That is not the way to dispose of the contents. Put the lid on and set it on the floor. We can dump it when we land." A moment later, I heard the wind roaring through the cabin from the open window on the passenger's side. McIsaac was in the right rear seat and he attempted to pour the contents against that wind. The wind blast carried the entire contents of that container directly back into Roddie's face and down the front of his jacket.

On another occasion, when I was in Wabowden, I was approached by an attractive young lady. In spite of the temperature, which was minus 22 Fahrenheit, she wore a skirt, a parka and silk or nylon stockings. She wanted to charter IXU to Cranberry Portage on Athapapuskow Lake. I loaded her baggage in the back seat and we departed Wabowden.

As we proceeded toward our destination, she would point at various lakes which we passed from time to time and ask if IXU could land in such a space. I wasn't exactly sure what she had in mind when she hoisted her skirt a few inches, exposing more of her well-shaped leg. I was never sure if she found me attractive or if she felt this might be a way around paying for air transportation. However, when we arrived at Cranberry Portage, she paid her fare in cash. I have often wondered if the lady ever questioned how a fellow as dumb as I must have appeared could ever have learned how to fly.

One afternoon, I had a charter from Flin Flon into Thicket Portage. I was to pick up three men who were moving from Moak Lake to Cook Lake. As I mentioned previously, Cook Lake is where Thompson now stands. I landed at Thicket Portage in decreasing visibility because of heavy snow. The snowfall continued and increased in intensity as the afternoon wore on. I knew I was stuck until at least the next morning. I was not alone; several other aircraft had also become snowbound. Every room in the hotel was booked. A good Samaritan offered his pool tables, at no charge, to anyone who had a bed roll. I returned to my Cessna 180 and came back with my Wood's Arctic three-star sleeping bag. Seven other pilots with similar bed rolls joined me; we spent a comfortable night sleeping on solid, green felt-covered tables.

During that winter of 1956-57, a man named Jim Wylie came to our base and chartered Cessna 180 CF-IXU for a flight to his hometown of White Fox. This Saskatchewan town is just a few miles northwest of Nipawin. I landed on skis in a field near the town; Jim took his pack sack from the rear seat to walk the quarter-mile into town. He would call us when he wanted to return to Flin Flon.

Since White Fox is quite near my hometown of Lac Vert, it seemed reasonable that I pay a short visit to my parents. I landed in my brother Murray's field and taxied directly to a spot opposite the driveway leading to my parents' farmhouse, almost exactly where I had seen my first airplane when I was a young lad. Murray happened to be there as well that day. Many years had passed since Murray had been a flying instructor with the Royal Canadian Air Force. But once a pilot has gained a certain level of proficiency, he never forgets how to fly. I had no hesitation in letting Murray try his hand at flying Cessna 180 CF-IXU. He did well, considering he had not flown for so long, especially since he had not been inside a Cessna 180 and had never flown a ski-equipped airplane.

My niece Dianne Williams (daughter of my brother, Jack) and several other nieces and nephews were over that day to see Gramma and Grampa. When I saw Dianne, I was reminded of those days in the mid-'50s when Louise and I would visit Lac Vert on vacation. Dianne would bring the other kids into the west room upstairs in my parents' house and I would tell them stories about flying in the north. They found the names of the various places intriguing: Athapapuskow, Pikwitonei, Wapawekka Lake, the Paskwachi River and Pukatawagan.

They all wanted to hear over and over about the day I was flagged down at Pukatawagan. I had been at Lynn Lake that day, returning to Flin Flon; our helper had refueled CF-IXU and I took off for a return flight. As I flew over Pukatawagan I saw many people waving their arms as though there was some emergency. I landed to see what was going on, and when I had docked, I was informed they had seen what they took to be white smoke trailing out behind my airplane. I discovered the helper had failed to secure the filler cap on the right wing tank. No use blaming the helper;

checking such things is the responsibility of the pilot. By neglecting to do the check, I had lost nearly half the fuel in my right-wing tank; the airflow over the wing had syphoned the aviation gasoline right out of the tank. People at Pukatawagan mistook the thin sheet of raw fuel as a curtain of smoke streaming out behind my Cessna 180.

There was no time for gathering in the west room today for more stories. My nieces and nephews were at my parents' house for lunch and had to return to school. I thought it might be interesting for them to arrive at school by air. Dianne, the oldest of the group, sat in the front passenger seat beside me and the others crowded into the rear seats. We took off from the field, and I landed in the field beside the schoolyard, about a mile from the farm. They disembarked, to the amazement of all the other school children who were playing in the yard. I took off again to fly back to the Flin Flon base.

A. R. (Al) Williams

7 Oscar Erickson's Belly

It was the winter of 1950-51, and from Pukatawagan in the north to Cumberland Lake in the south, and from Deschambault Lake in the west all the way to Cross Lake in the east, several inches of new snow had fallen during the night. This winter had been harsh. Strong winds had drifted the heavy snowfall into ridges in unprotected areas, which were as hard as concrete. Oscar Erickson had a charter flight this morning to fly mail and groceries plus several barrels of gasoline for the power generator to the Sawmill on the southeast shore of Reed Lake. Since this flight would take him near old Sven Bjorenson's cabin on the north shore, Oscar picked up mail for the old trapper. There was a copy of the latest Eaton's catalogue, a letter from Sweden and an envelope from the Hudson's Bay Company, which might, for all Oscar knew, contain a check for Sven's last fur shipment.

Oscar looked toward Flin Flon, three miles north, and observed that the wind was virtually calm. Sulfur-laden smoke from the smelter's smoke stack drifted lazily in a windless sky. A pilot flying downwind from that smoke stack, with poor visibility, was often able to follow the sulfur taste at the back in his throat all the way home.

Oscar readied Norseman CF-GTP for flight. As he had done so many times before, he took his blow pot and fire extinguisher under the engine cover. The blow pot was nothing more than a gasoline-burning blow torch to warm the engine for starting in winter. Oscar first observed the dark metal of the engine turn brilliant white with frost. He waited until that original dark color returned to tell him that the frost had melted and the temperature had risen so the engine could be started. Although the temperature had risen somewhat on this particular morning, the dull gray overcast sky made it difficult for him to see the surface of the newly fallen, unbroken snow.

During his takeoff run, Oscar's flat-nosed Norseman disappeared behind the heavy curtain of snow. The snowy barrier was caused by slip stream from the three blades of his Hamilton-Standard propeller driven to 2,250 rpm by the 600-horsepower Pratt & Whitney R-1340 wasp engine, with prop tips traveling faster than the speed of sound. GTP became visible again only as she climbed out over the lake and the snow settled back to the surface once again. The flight to the sawmill was uneventful. Even the landing was made simple because the crew knew Oscar was coming. Being aware of poor visibility under unbroken snow and overcast conditions, the men had marked out a runway with spruce boughs along a stretch of lake ice that was fairly free of concrete-hard drifts.

Oscar's load was removed from GTP, and he was invited to a hearty meal prepared from his cargo. Roast beef, potatoes and gravy with fresh vegetables. There was even a choice of pies, plum pudding or carrot cake for dessert. Having eaten and after sharing the latest news from outside, Oscar was airborne again as he headed northwest on the course to Sven Bjorenson's cabin.

Because of the hard drifts, invisible now beneath the unbroken snow and overcast sky, Erickson set up his landing approach to put GTP on the ice surface behind the lee of an island just outside of the bay in which Bjorenson's cabin was located. Oscar knew that the prevailing winds, sweeping south eastward along the northern shore, would not have had much chance to form hard drifts near the lee shore of the island. His landing was quite smooth, and he taxied around the end of the island and headed into the bay. A long, shallow ramp of snow had drifted directly in front of the cabin, about 100 feet out from shore. There was a sharp drop-off on the cabin side of the drift. Fresh snow that had fallen during the night under a very gentle wind had filled the steep valley. Oscar could not see what lay just beyond the tips of his Elliott skis. The angle was shallow enough that Oscar was scarcely aware that GTP had mounted a ramp.

As his skis fell abruptly to the level of the lake's ice, Oscar heard the snapping sounds of breaking "stringers" and the creaking of twisted "formers." He had taken the belly out of GTP on the crest of that rock-hard snowdrift. The fresh snow was nearly four feet deep on the valley side of the drift, and Oscar had to apply a considerable amount of power to free his airplane's belly from the snowbank.

Oscar was a pilot, not an engineer. While he understood that a few broken formers and stringers wouldn't cause a reduction in the structural strength of the airframe, he didn't know what would happen if he flew

GTP with those belly formers and stringers removed. He decided not to find out.

Once the engine cover was in place, Oscar went to the trapper's cabin for help, but Sven was out on his trapline and it could be many days before his return. The log cabin, as was common in the north country, was not locked. The pilot lit a fire and put a kettle of water on for tea.

He removed his ax from a compartment aft of the left cabin door of GTP and walked through knee-deep snow into the forest in search of some lively willow branches. One by one, he removed the broken and twisted formers to replace them with carefully measured willow branches. Estimating the right shape and size, he secured them in position with fishing line salvaged from Sven's storehouse. He stopped only once for tea in late afternoon. When he quit work at sundown, most of the formers were in place. Oscar hoped that his makeshift repairs gave GTP a normal bulge to her belly.

No word of GTP had been heard since Oscar had left Flin Flon in the morning around 10:00 hours. By 19:00 hours, radio sked time, questions were asked. "All stations, Flin Flon ... has anybody heard GTP today?"

"Nothing heard here, Lynn Lake." No one had seen or heard from Oscar, but suddenly his voice came over the radio.

"Flin Flon, GTP. I've been delayed. I'll spend the night at Reed Lake; would you call my wife and let her know?" Oscar's radio message was a relief. At Sven's cabin that evening, he made supper and then settled down to sleep in his Arctic bedroll on Bjorenson's bunk bed.

Next morning, Oscar was hard at work. He had installed all the formers and began building stringers, which would run longitudinally to support those formers. The stringers were made of thin willow branches. He bound them together with more fishline to make them long enough; these, in turn, were tied to the formers so they would retain their strength as belly fabric was stretched in place. The problem was that fabric which had covered the belly was shredded. What could be used for a covering?

Erickson pondered this question for some time. He found a bed sheet in the cabin, but what could be used for dope to glue the sheet? As the temperature was now minus 30 degrees Fahrenheit, he wondered if the bed sheet could be frozen in on the plane's belly.

Oscar used his ax to chop a hole in the lake's ice surface and filled a cooking pot with water. He took a bed sheet and laid it out on the ice beneath the belly of GTP. He began holding the sheet in place a spot at a time, as he splashed a small dab of water at each spot. It was painstaking work, but it wasn't long before Oscar had stretched the sheet over the rough framework of formers and stringers until old GTP looked almost like new. He continued dabbing water along the seams until he was satisfied that it was secure; then he took his blow pot and began to warm up his engine.

Oscar made some rough calculations (based largely on feel, as he had no weigh scale) that the weight of the willow branches would match that of the old, original formers and stringers. He taxied GTP to the lee of the island under a pale sun, which greatly improved his visibility. The pilot applied near-takeoff power, but only long enough to reach fifty miles per hour. Then he brought the GTP to a stop and climbed down to examine his handiwork. The ice dope held and there seemed no reason to delay takeoff. GTP and Oscar winged their way safely back to Flin Flon.

Today, with all-metal aircraft, such field repairs might offer more of a challenge. This does not, however, detract from the inventiveness of Oscar's solution. CF-GTP spent thousands of hours more in the air after permanent repairs. A few weeks after his incident, Oscar again dropped in with mail to visit his old friend Sven Bjorenson. He replaced the missing bed sheet and restocked Sven's pantry. He picked up another load of furs as well. The two men had a good tongue wag about Oscar Erickson's belly.

Oscar Erickson was also the pilot during a dramatic event in 1953, about three years after his Reed Lake adventure. An Indian Affairs doctor was concerned about an Indigenous woman, Mrs. Ballentyne, who was almost nine months pregnant. She'd experienced complications during her previous pregnancy, and the doctor was concerned those problems could recur during the second birth.

Mrs. Ballentyne lived at Sandy Bay, about seventy miles northwest of Flin Flon. Her doctor requisitioned air transport whenever he believed an Indigenous patient's condition warranted a hospital stay. January temperatures had dropped to the low minus 40s when Mrs. Ballentyne's physician decided that his patient should be in hospital during her final two weeks. Central Northern Airways' base, with Oscar as pilot, was chartered to fly the patient from Sandy Bay. Oscar readied his Norseman for flight, and in due time the big radial Pratt & Whitney R-1340 wasp engine in CF-GTP was ready to run. The agent and his helper loaded blankets and a pillow onboard so the patient could recline during the return flight to Flin Flon.

Oscar taxied out onto Schist Lake, becoming almost invisible as a cloud of fresh snow billowed up again from the surface under the influence of those three blades of GTP's propeller; he turned to face the cold north wind and applied takeoff power. His takeoff run was quite short. Because of dense air and wind, he climbed over the shoreline with much more altitude than he would normally have in warmer, less windy conditions.

The flight from Schist Lake to Sandy Bay was quite uneventful. As Oscar turned on his final approach to land at the settlement, a group gathered in anticipation of his arrival. They fetched the pregnant Ballentyne for her evacuation to the hospital. Initially, she elected to sit in the passenger seat to the right of Oscar. They took off and turned on the heading which would carry them back to Channing on Schist Lake's northern shore, for her transfer to Flin Flon Hospital. After they had been airborne for about twenty minutes, Mrs. Ballentyne went into labor; she asked that Oscar land at once to give her whatever help he was able to give. Oscar looked for a suitable landing surface and put down on a small lake just left of their flight path. Mrs. Ballentyne climbed back into the cabin of the Norseman. An inexperienced, anxious Oscar acted as midwife, birth coach and doctor during the birth of her baby boy that, blessedly, didn't result in complications. Oscar helped Mrs. Ballentyne into his own Wood's bedroll for the remainder of the flight. An ambulance summoned by his radio was standing by when GTP arrived back at Schist Lake; Mrs. Ballentyne and her new son were transported to the hospital.

Oscar's monthly base pay plus mileage flown was the only reward he expected for this flight. He wasn't expecting Mrs. Ballentyne's bonus—the newborn boy was named Oscar Norseman Ballentyne.

* * *

Notwithstanding the rigors of hard drifts and the difficulty of landing on unbroken snow under an overcast sky, I personally enjoyed winter flying. I liked the fact that during winter it was not necessary to tie ropes to the airplane to keep it from drifting away. During daylight hours it was not necessary to insulate the skis from a snowy surface; they only froze overnight unless you parked where water could seep out from under the snow.

Another thing I liked about winter flying, during unbroken snow conditions, was flying a Cessna 180 and using the HF radio as an indicator of contact with the surface. Unbroken snow is much like glassy water—under an overcast sky, snow surfaces become invisible. These landings are essentially made by instrument only.

An aircraft's skis are rigged so the ski nose is a bit higher than the heel during level flight. A rubber bungy holds the ski nose high and keeps a short steel cable taut between the heel and fuselage. In Cessna 180 CF-IXU I always kept the high-frequency communications radio on. When approaching to land on unbroken snow, the heels of my skis would touch and the overhead loudspeaker made a "kitsch" noise at the instant they made contact with the snow. This is similar to the curb feelers many people installed on their cars in the 1950s and 1960s. The noise in the speaker originated in discharges of static electricity on the airframe at the instant of contact, similar to atmospherics in a high-frequency radio.

8 Other Adventures

Nature doesn't always cooperate with flying people, and I remember a flight that provided more than a few too many seconds of excitement.

I had been hired to fly a prospector into a very small lake not far east of Wekusko, east of Snow Lake, in northern Manitoba during 1957. The lake had no name, but because it was shaped like a mouse, complete with a tail that took the form of a creek running away from the south end, we called it Mouse Lake.

The approaches were very flat, with only muskeg extending more than a half mile north and south. It was easily accessible, and it was possible to approach with no more than ten feet of altitude for several hundred yards, from either direction. This lake was perhaps 600 feet long and about 200 feet wide.

I came in low and slow from the north end with 40 degrees of flap extended and placed IXU's skis on the surface within only six feet of the north end of the surface. I hadn't realized until that moment that the surface was glare ice; the wind had swept it clear of snow. Even though I pulled the throttle all the way back to idle, we didn't seem to be slowing down quickly enough. A few rocks were scattered across the south end of the lake, only a foot or two above the ice. Those rocks could do a lot of damage if I hit them with any speed. When it seemed absolutely certain that I was going to plow into the rocks, I hit full right rudder and opened the throttle to induce a controlled groundloop. IXU came smartly around 180 degrees. I chopped power once more. We slid backward about forty feet after IXU had turned around.

During takeoff from that lake, the smooth surface allowed for very quick acceleration and we passed happily over the end of the lake at more than fifty feet above those lethal rocks.

This story reminds me of the day my brother Jack was out on an exercise with his Royal Canadian Air Force flying instructor, Jack Park.

Jack Park was Australian. He had left his home in Queensland to join the Royal Australian Air Force for elementary flight training. Park was familiar with airplanes because of weekly visits by aircraft landing on the strip at the sheep station that his father managed.

At twenty, Park accepted a transfer to Canada to continue his advanced flying training. He became an RCAF sergeant in the British Commonwealth Air Training Plan and was posted to a training school in Brandon. A fine instructor, he turned out many polished young RCAF pilots, who were desperately needed in the European theater of WWII. Park passed his enthusiasm and a good deal of his skills along to his students.

One of Park's students was a young Canadian pilot, L.A.C. Jack Williams. Williams had moved to service flying training school, in which training was done in twin-engine Cessna Crane aircraft.

On a cold January 18, 1941, Sergeant Park had a few of his students with him on a training mission. They had been practicing low flying. When the lesson was over, Park took control and announced his intention of showing them some real low flying.

The aircraft, serial number 7763, was one of the newest models of Cessna Crane. Its Jacobs L-4 MB engines were fitted with the new constant-speed metal propellers as compared with the fixed-pitch wooden "clubs" that were standard on earlier models.

Sergeant Park came over the shoreline of a lake near the practice area and began to descend. Their apparent speed increased with every inch their Cessna Crane descended toward the lake's frozen surface. When it seemed impossible to fly any lower, Sergeant Park allowed his Cessna to inch just a bit closer to the ice, with its left wing slightly low.

Suddenly those on board heard and felt a machine-gunlike rattle. The left propeller had contacted the icy surface!

He pulled up abruptly; and as he cleared the trees, a flock of partridge flew directly into his path—one bird coming to a violent end as it smashed through the leading edge of the wing and lodged against the main spar.

Park was unable to climb above 100 feet because of the low clouds and flew directly back to the relief airdrome. The aircraft shook strongly, giving new meaning to the nickname most pilots had given the Jacobs engines—"Shaky Jakes."

All occupants aboard 7763 breathed a sigh of relief when their wheels rolled smoothly onto the centerline of the runway. Once back on the ground, Park taxied to the tarmac. Examination of the propellers revealed two inches of the left tips folded backward. Pilots in the mess hall jokingly said that Park normally had superb judgment and had flown to a perfectly safe altitude for wooden propellers, but he had forgotten this particular aircraft had been fitted with those new, longer metal props. It was nonsense, of course, but it indicated the confidence the other pilots placed in Jack's usual depth perception!

If they were asked, the students were to say that the oleo leg on the left landing gear had gone flat during the wheel landing; thus the damaged propeller. It was academic, however, as they were never asked.

Two days later, L.A.C. Williams flew 7763. She performed flawlessly—a tribute to the good workmanship by the aero mechanics and fitters.

* * *

In the spring of 1957, about nine years before I worked with Bob Lundberg, I was approached by U.S. (Ulysses Stanley) Wagner, assistant general manager of TransAir. He asked me to join the company as pilot at the Flin Flon base. Wagner was persuasive. "You seem to be stuck flying Cessnas and Stinsons, but if you come with us you will soon be flying Norseman and Beavers."

Stan and I had felt a certain antagonism for each other since my stint as district radio operator for the Department of Lands and Forests at Sioux Lookout in the early '50s. Wagner had then been the base manager for Central Northern Airways at Sioux Lookout. This was prior to Central Northern joining Arctic Wings to become TransAir.

Stan was then in the habit of turning up the modulation control on the radio transmitter at his base, to the point where he was causing great interference to our communications system. Ross Keenan, radio technician for Lands and Forests, had visited Central Northern's base at least twice, taking an oscilloscope for testing. Keenan found that the transmitter was grossly over-modulated. Over-modulating a radio frequency carrier produces a great deal of distortion and generates heavy harmonics of the audio signal. The signal occupies much more spectrum than is needed. The signal is also much more difficult for the person receiving to understand, as other central northern operators can attest. Keenan showed Stan the effects of over-modulation on the scope.

Ross Keenan proved that Stan was abusing the radio. Maybe Wagner didn't like being corrected by a mere mortal. Whatever the reason for his lack of cooperation, one day we locked horns over the interference.

As mentioned, I was Lands and Forests' district radio operator responsible for overseeing all radio communications circuits in the Sioux Lookout fire district. Our boss, Harry Middleton, held the title of district forester. Jim Rorke was forest protection supervisor. Jim was second-in-command of the Sioux Lookout fire district. The chief ranger at Red Lake was Frank Dodds. The chief ranger in Pickle Lake was Gene Guertin. Each subdistrict, working under Harry Middleton, had a radio operator. Howard Knapp was the operator at Red Lake; Jo Harvey was at Pickle Lake. Jo was at her post the day this story took place.

On this particular August day, Ms. Harvey came to work, but she was not feeling very well. Before long she found it necessary to visit the nursing station. The nursing staff decided Jo had a ruptured appendix, and since it was only a nursing station, staff advised Gene Guertin of Ms. Harvey's condition and told him she would require transportation to the Sioux Lookout hospital for surgery.

Harry Middleton was at a three-day district forester's seminar located at some out-of-the-way retreat in the Kenora District, with no means of being reached. At the same time, our forest protection supervisor, Jim Rorke, was away on urgent business and could not be reached for two days. It was imperative that Jo Harvey be transported at once.

Chief Ranger Gene Guertin was approaching retirement, and while he was a very kind individual, he was afraid of doing anything that might place his retirement in jeopardy.

As chief, Gene could requisition an aircraft for a flight within the confines of his own subdivision, but he normally needed authorization when a flight outside his subdistrict was required. On this particular day, Gene Guertin sent a radio message to be delivered to the district forester. The chief ranger's message was addressed to Sioux Lookout; I did my very best to copy his message through severe interference from Central Northern Airways' overmodulated transmitter signal. Gene Guertin's radio message was addressed to district forester Middleton and transmitted as follows:

A. R. (Al) Williams

> *Nursing station advises radio operator Jo Harvey has ruptured appendix X request permission transport to Sioux Lookout by Beaver CF-OCS X urgent.*
> Gene Guertin, Chief Ranger

During several attempts to receive this message, Stan Wagner's over-modulated signal came on the adjacent frequency and made it impossible for me to copy the message. The few words I was able to understand alerted me to the fact that we were in a life and death situation. At this point I picked up the telephone and called Wagner. "Get that damn transmitter off the air, Stan. I've got an emergency at Pickle Lake, a matter of somebody's life!"

"Don't you swear at me, goddamn it!" Wagner snapped at me and slammed the telephone down.

I immediately called him back and said, "This is the last time I have any intention of talking to you. I'm calling the radio Inspector in Fort William, and we'll let him decide on what's to be done about your interference." I went back to Gene at Pickle Lake and his message, but I was angry over the total lack of sensitivity Wagner had shown over this life-threatening situation.

As I received the message, I had to advise Guertin that it was not possible to deliver his message because the district foresters were at a retreat. Guertin asked that I deliver the message to our forest protection supervisor, and again, I had to advise him that Jim was also incommunicado. I added, "There isn't any choice, Gene; you have to get Jo Harvey to the hospital."

Gene's reply shook me up. "I can't requisition this flight without authorization."

I thought it over for about four seconds, knowing that the nurses at Pickle Lake must surely know what they were talking about. I pressed my foot down on the foot switch and said, "Well, Gene, Jo must be brought to Sioux Lookout, will you accept this authorization from me?" (As district operator I had absolutely no authority whatever over anything having to do with the overall operation of the Sioux Lookout Fire District.)

"Yes, I'll accept authorization wherever I can get it," he replied.

I typed out a brief message to Guertin: "You are hereby authorized immediately to transport Jo Harvey via CF-OCS to Sioux Lookout Hospital" and signed my name.

Within five minutes, de Havilland Beaver CF-OCS flown by James

Kirk was on its way, bringing Jo Harvey to safety. Jim Rorke was delayed and didn't return until Monday about noon, but Harry Middleton was in his office at 08:00 hours the same Monday. I telephoned Harry Middleton at 08:10 hours advising him of the happenings: I told Harry that, right or wrong, I would do the same again under the same circumstances.

Rather than criticizing me, he patted me on the back and praised me for showing initiative in a life-or-death situation. He was less complimentary of Gene Guertin, because he said Gene was aware, or should have been, that in an emergency the rule book was always tossed aside and common sense was expected to prevail.

Jo Harvey survived her surgery; in three weeks she was back doing her job at Pickle Lake. In due course, Gene Guertin took his retirement, without any fallout from the emergency flight. I believe there would have been no fallout if he had thrown the rule book aside and ordered the flight for Jo Harvey. Harry Middleton later said that he had received a telephone call from Stan Wagner, complaining that I had sworn at him on the telephone. I asked Harry if Wagner had told him about the circumstances. Middleton then said that it was a closed issue.

A few days later, Joe Stevens, Radio Inspector with the Department of Transport Radio Inspection Branch, paid a visit to Wagner and laid down the law. Central Northern Airways would not, from this day onward, under any circumstances, interfere with Department of Lands and Forests, or any other local radio communications installation. If another case of interference was reported to the Department of Transport radio inspector's office, Central Northern Airways would have no alternative but to hire a fully licensed radio operator. Central Northern Airways was not about to waste their money on what they considered unnecessary help in the form of a licensed radio operator. We had no further problems, but Wagner remained angry at having been faced down by a lowly civil servant like me, especially one of about twenty-three years of age.

9 New Experiences

I was surprised that Stan Wagner at TransAir wanted to hire me as a pilot due to our former dispute over the radio interference. However, the challenge of Beaver and Norseman flying overcame my caution. I would have reason to regret this decision, but at the same time it provided an opportunity to learn many new aspects of aviation.

Soon after I had joined TransAir, former customers Archie Talbot and Ingy Bjorenson came to charter my services. Archie and Hank Parsons (my former boss) were close friends, but Archie wanted to support me as well. Archie told me that Parsons had been complaining about pilots who received training at his expense and then drifted off to fly for competitive companies. Parsons had also admitted to Archie that I was a good, reliable pilot who had always done my job well. I could understand Parsons' gripe: when you have a good crew that works well together, as we did, it throws a wrench into the works when someone leaves and you have to hire a new team member.

At the same time, I knew that Hank Parsons was not a man who wanted to hold anyone back from career advancement. He was grumbling because business was falling off a bit and some of my former satisfied customers had brought their charters to my new employer.

Apart from the fact I was not logging the number of flight hours at TransAir that I had been accustomed to at Parsons Airways Northern, I was happy with the prospect of flying Norseman aircraft. Robert Bernard Cornelius Noorduyn originated the design in 1934. His first production prototype airplane was serial number one and was test flown November 14, 1935. That aircraft was designated as a Norseman Mark I and was registered as CF-AYO. CF-AYO was powered by a 420-horsepower Wright engine. It was the only Mark I Norseman ever built. In August 1953, CF-AYO was destroyed at Round Island Lake, Ontario.

In 1936, production of Norseman Mark 11 began, but the airplane proved to be underpowered. Only three aircraft of the Mark II series were built: CF-ASA, CF-AZE and CF-AZS. Just one Mark III Norseman, CF-BAM, was built, powered by the Pratt & Whitney C Wasp 450-horsepower engine. Though somewhat better performance was realized, the aircraft was still underpowered.

The prototype Mark IV Norseman, CF-BAU was powered by a Pratt & Whitney H Wasp engine rated at 600 horsepower. The first production Mark VI Norseman was serial number 100; a total of 900 Norseman aircraft (all Mark numbers) had been built when Mark VI Norseman production ceased with CF-GOB September 7, 1951. The Mark VI was also powered by a Pratt & Whitney R-1340 Wasp engine rated at 600 horsepower. Virtually all Norseman craft I was fortunate to have flown were Mark VI.

The Mark I Norseman (serial number one) CF-AYO was used in a 1942 Hollywood movie *Captains of the Clouds*, starring James Cagney. Canadian World War I Ace Billy Bishop also appeared in one of movie's scenes that was shot in Canada. The fuselage of CF-AYO was put on display in the Bushplane Heritage Museum, Sault Ste. Marie, in 1994.

During the summer of 1957 I wasn't all that busy, but I did make some interesting trips, sometimes accompanied by Ron Dodds, a helper employed by TransAir on Schist Lake. Ron refueled my aircraft and helped with loading and unloading the Norseman, especially where we had to move a client's camp from one lake to another. He was a delight to work with—always bright, cheerful and willing to lend a hand without being asked. We flew together quite a lot that summer. Once we were airborne and on course, we would often change seats and Ron would maintain our heading, making a few gentle turns. When our destination came into view, we would change back for the actual landing. I knew that I could easily override Ron at the controls because of the open structure under the instrument panel and the position of the control column between and just forward of the flight deck seats. However, Ron was a quick learner and I never had to take over the controls. He had natural ability to become a competent pilot.

One day in August 1957, we were called upon to move a Midwest Diamond drill camp from its current location to another site just two miles to the east, in the Hanson Lake area. The drilling camp often moved itself between locations with ground transport, but in this case, the area between camp locations was very rough. Large outcroppings of rock blocked their John Deere track tractor in all directions.

Ron Dodds and I arrived early that morning to fly the camp to a new location because there were to be many trips and much loading and unloading to be done. We packed the camp up quickly while the crew began to disassemble the John Deere tractor. On our next trip, we moved the tractor's frame and tracks. Then came the John Deere engine and some of the drill equipment. Finally we moved the transmission.

I was amazed how quickly the crew could disassemble and reassemble that tractor.

The wind was blowing from the north all day, and since the camp was on the northwestern corner of the lake, I needed to taxi quite a good distance to secure enough takeoff room. The new campsite was on the southeastern corner of its lake, so again I had to taxi a long way at the new location. All in all, we made forty takeoffs and landings to moving the camp just over two miles. Finally evening came; our job of moving camp was complete. Ronnie and I were free to fly back to Flin Flon after a very tough day's work.

"Well, Ronnie," I said, "Tonight's the night you're going to do your first takeoff." Ronnie stared at me as though I had lost my marbles. "Now listen, Ron," I said. "You've flown this Norseman enough to know the way she feels in the air; this is really no different. You know the power settings—thirty-six inches and 2,250 for takeoff—thirty inches and 2,000 for climb— twenty-seven inches and 1,850 rpm for cruise. You know you can do it." A bit reluctantly, he took the left seat and we taxied out onto the lake. He turned into wind, raised the water rudders and began his takeoff run toward the hill at the north end of the lake. Everything was going smoothly. Ron had pulled the throttle back to a bit less than thirty inches before reducing the engine rpm— to allow for an increase in manifold pressure when climb power was established. Dodds then turned the flap crank four turns to reduce flaps for best climb.

We were about 200 feet above the trees atop a hill when the left cockpit window exploded. Cabin noise instantly increased at least 80 decibels and Ronnie looked shocked. "Don't worry, Ronnie," I yelled, over the roar. "You're doing fine. It's just the window. Fly the airplane—don't let a little thing like that put you off." Although obviously shaken, Ron didn't panic and continued to fly. I am absolutely certain I could have regained control of the airplane if he had panicked, but the only input needed was calm words of encouragement.

When we reached our cruising altitude of 2,000 feet, a still shaken Ronnie wanted to change seats with me. I convinced him to just relax and he flew us to within sight of Schist Lake; at that point I took over, flying GTP to a normal landing. We never did discuss that flight very much, and we never discovered any reason for that window to explode. I felt a great deal of pride in my young helper/student pilot.

Following the job of moving the diamond drill camp, our flying activities at TransAir slowed down considerably. We would get the odd trip to the Sandy Bay settlement with express and three or four passengers but, in the main, things were quiet. It was understandable, then, that I was overjoyed when someone requested me for the job of flying an entire school roster of children from a residential school to their homes for summer vacation. Stan Elliott, our Wabowden base manager had contracts for a fish haul and could not get away from his commitment for the yearly transportation of Cree children from the Island Lake residential school. Elliott was anxious that I move to Wabowden, temporarily, to complete the student transfer contract.

In June 1957, I flew Norseman CF-GTP to Wabowden, where I could refuel and have engineering checks done. I knew that I would need an adult, preferably an Indigenous man, to act as supervisor when the children were onboard Norseman CF-GTP. I asked Stan Elliott if Mark Rabbit was available. Rabbit had built a beautifully crafted log house, which he shared with his mother.

His mother, a superb cook, kept the home spotless. She was justifiably proud that Mark was employed part-time by TransAir. Mark was an excellent crewman in addition to being a very pleasant person. I was glad to learn that he was available for the school flight. After I delivered a load of fish for Stan Elliot, Mark arrived at the dock with his mother's kind invitation to join them for dinner.

Early the next morning, Mark and I were off to Island Lake; we made several trips to ferry children to their homes. When we had finished flying for the day, both of us were very tired and looking forward to a good night's sleep before resuming the charter in the morning. A local priest told me I would stay in a rectory guest room. "Your man will bunk in with the Indians," he casually added. My philosophy is "the customer is always right" (unless he is completely wrong). I knew I had to take a stand against the slur aimed at my friend and helper, Mark Rabbit.

"No," 1 said, "Where any member of my crew stays, that is where I stay." The priest looked at me as though I had taken leave of my senses. "You can't sleep there," he told me. "It is far too dirty."

I looked the priest in the eye. "If it's too dirty for me, it's too dirty for my crew. You'll have to do better than that." I didn't want to argue with a customer, but I was adamant.

"Well, you see," he said, "it just isn't done."

"Fine. Mark and I will sleep in the airplane," I said. Then I added: "This Jesus you pretend to follow would be turned away if he came here seeking sleep. His beard would need trimming or his hair would be too long, or perhaps his clothing would be unsuitable for your accommodations. So much for your Christian principles."

I turned and began walking away, but the priest called me back. "I apologize," he said. "You're quite right. One of the brothers will fix up the other guest room for your man."

Every morning, for the balance of our stay, Mark and I awoke refreshed and ready for another day of flying. Nothing more was ever said about the priest's attitude or my stand, but all future stops at Island Lake Mission were handled to my complete satisfaction. I wonder, today, in light of the known physical and mental abuse that so many residential schoolchildren suffered, about the cruel treatment those Island Lake residential school kids may have been subjected to. At the time, the idea of the clergy abusing kids simply never entered my mind.

When I took over the position of base manager at the Flin Flon TransAir base on Schist Lake at Channing in spring 1958, we had only about 3 percent of flying business out of Flin Flon. Parsons Airways Northern Ltd. maintained a tight grip on the remaining 97 percent. One reason for this disparity in our market share was based on TransAir's history of being unreliable; for instance, a customer was promised a move from his current camp location to a different lake. His move was to take place at 10:00 hours on Wednesday, but the aircraft failed to arrive until 16:40 hours the following day. Needless to say, the customer was most unhappy. He'd broken camp for a quick transfer, only to find it necessary to at least partially reassemble his camp again for a night's stay.

A reputation for unreliability in any business is difficult to change, no matter how impressive the company logo or head office's mahogany paneling might be. In the aviation business, the base manager usually sets the tone for the conduct of business. As a lowly line pilot, before I was promoted to TransAir base manager at Channing, there was not much I could do to influence our effectiveness. My predecessor was a base manager named Neville Donahue "Hoppy" Hopson. Originally from Texas, where everything is reputed to be larger than life, Hoppy was just five-foot-two and weighed 108 pounds. He possessed a sharp tongue and was quick to disagree and criticize.

A customer walked through our front door after we hadn't turned a propeller for nearly a week. "Mr. Hopson?" the man said, "My name is Mathews—Jerry Mathews. I am with Noranda Mines, and we're doing some exploration work in the Deschambeault Lake area. We have about thirteen tons of men and material to fly in there." He paused to looked out at our Norseman tied to our dock. "Since our camp will be about eighty miles from here, we expect you can make three trips per day for five days and move all our stuff in there. Can you see any problems doing that job for us?"

Hoppy went to our credit-list book, although that was unnecessary since everyone knew Noranda Mines was good for any fee that an employee negotiated with any local business. It would have been a mere formality to sign them up if Hoppy had simply handed the man a credit application sheet. While Hoppy was checking the credit list, I was mentally adding up mileage and calculating my flying pay. At four cents a mile, the job would pay four dollars per hundred, with each round trip of 160 miles; why, that would make 2,400 miles at four dollars per hundred— I should earn flight pay of nearly $100. I was elated—but not for long.

"Mr. Mathews," Hoppy began, "I don't see a listing in my credit book." He paused for a second and then went on, "Who's going to pay for the charter on this much flying? I'm not going to be stuck with a penny of it!"

Mr. Mathews was taken aback but did not give an inch. "Mr. Hopson, I agree with you; you shouldn't be stuck with the charges ... I am sure I can make other arrangements." With that, Jerry Mathews left. Hopson realized his mistake, but it was too late; the man disappeared in the taxi he had waiting.

The taxi went directly to Parsons Airways, where he was greeted by Hank Parsons. Parsons told him he would have a Norseman available at his dock at 08:00 the next morning. Mathews departed to begin organizing the loads that Parsons would fly out of Schist lake. Parsons didn't have a Norseman. However, as soon as Jerry Mathews left, he phoned Tom Lamb of Lamb Airways at The Pas. Hank Parsons chartered a Norseman for five days—billing was to be made directly to Parsons Airways Northern Ltd. The contract went without a glitch, and Noranda Mines named Parsons Airways for any future work in the area. It had cost Parsons about $100 more than he charged Noranda, because he paid for the ferry flight from The Pas and return, but he was now an established supplier for Noranda. As it turned out, Parsons made a profitable investment that day, because when Noranda Mines continued their work, they gave him a lucrative resupply contract. That contract could have been ours had Hopson been on the ball. Although 1 liked Hoppy personally, I told him that one of us had to leave. Finally, Neville Donahue Hopson accepted a position in our payroll office, where his record-keeping skills would be appreciated and where there could be no further contact between him and our customers.

I was named base manager and spent a great deal of effort trying to win back disgruntled customers. I would visit them at their town offices and fly out to the various fish camps.

For several weeks during the late summer of 1957, before 1 became base manager, I had been experiencing stomach and chest pains at night after I went to bed. During the day I felt fine, and I put it down to a case of indigestion. At night I could forget the discomfort by operating my amateur radio station until 03:00 hours, when I could sleep. When the pain continued, Louise insisted that I see our doctor. His test results indicated gallstones. Dr. Snyder said that to relieve my pain, surgery was necessary. He asked me when I was available. "How about tomorrow morning?" I asked.

It was settled. Next morning, on October 8, 1957, exactly eight years to the day after Louise and I had met (on a Saskatchewan government transportation bus) and exactly one year to the day before the arrival of our son Mark, I underwent surgery for the removal of my gallbladder.

We had been with the Manitoba Blue Cross medical plan for over a year at the time of my surgery, and the bill was forwarded to their

Winnipeg office. Within a few days, I received a letter from Blue Cross that stated the board had concluded that my gallstones were a "preexisting condition, therefore, we cannot accept responsibility for the (surgery) debt incurred."

The following day, I went to the Grey Nuns Hospital in Flin Flon and questioned three doctors about the cause of gallstones and the length of time it takes for the condition to develop. Each doctor told me that they didn't know precisely what causes gallstones or how long the stones take to form.

I banged out a letter on our new typewriter to the medical board, informing its members that my extensive research had revealed that medical science did not know the cause of gallstones nor the length of time necessary for them to form. "You will, therefore, immediately forward a cheque in the amount of $562.30 to cover the medical expenses incurred by this surgery." Within seven days I received a letter and a check. The letter stated that the medical board had reconsidered its original opinion and decided that my gallstones were not a preexisting condition.

On October 20, I invited Doctor Snyder to ride with me while I made a test flight on Norseman CF-GTP. He was not all that enthused, but I pointed out that I had placed my life in his hands while he practiced his profession. It seemed only fair that he reciprocate.

Our test flight went off without a hitch (outside my inability to raise the water rudders without the doctor's physical assistance). However, I was now informed by the Department of Transport that my license was suspended for about six months to allow me to recover from surgery. I was again in need of a job. Fortunately, the TransAir radio operator/agent at the Flin Flon Airport had just handed in his resignation. Head office knew that I had been an operator/agent with Canadian Pacific Airlines Ltd. from 1949 until 1951 and offered me the vacant operator/agent job.

About two years before I joined TransAir, it had taken over Canadian Pacific Airlines (CPA) operations out of Winnipeg. CPA radio operators who stayed on were critical of many of the changes the new company had brought in.

In retrospect, I can almost hear the moans and groans of those operator/agents when they heard that a bush pilot was coming to the Flin Flon Operation.

Since I had continued to operate my amateur radio station, for the most part using Morse code, I was not rusty. My years with CPA had honed my Morse speed. The other operators were pleasantly surprised when I took over the Flin Flon station. While CPA's Morse circuits were reputed to be among the fastest anywhere in the world, TransAir's operations were somewhat slower.

My pilot experience had trained me to prepare my weather reports (ceiling, visibility, temperature and dew point with relative humidity, estimating cloud types and heights).

The company had recently initiated a series of fisherman specials. These flights operated Tuesdays and Thursdays, to increase traffic from Winnipeg to Flin Flon and on to Lynn Lake during fishing season. They proved less than successful. Very often the DC-3 would arrive at Flin Flon and deplane three or four passengers to continue to Lynn Lake as an empty airplane.

The previous year, when I was flying Norseman CF-GTP, the regularly scheduled TransAir DC-3 would operate to Lynn Lake, except on Tuesdays and Thursdays, when it was terminated at Flin Flon as flight 87. As flight 88, it returned to Winnipeg via The Pas and Dauphin. At that time the DC-3 would remain at Flin Flon for about three hours before returning to Winnipeg. It had sometimes been possible for me to invite Captain Bob Stanley to fly my Norseman. Bob had been a pilot for Arctic Wings before the merger that had created TransAir, and I had every confidence in his ability to fly my Norseman.

Now that I was working as operator/agent, Bob Stanley would return my favor. On those days when we operated an empty DC-3 to Lynn Lake, Stanley would invite me to take the left seat of the DC-3 while he acted as first officer. The regular first officer would take a jump seat and I would act as pilot-in-command from Flin Flon to Lynn Lake. I would fly the DC-3 on the return flight unless we picked up passengers, but usually we returned empty. I was not able to log this flight time because I was operating under a medical suspension, but nonetheless, I had the experience of acting as pilot-in-command of a DC-3 for a total of perhaps ten hours.

Had I been able to fly a DC-3 anywhere near the hours I had flown in the Norseman, there is a good possibility that the DC-3 might have replaced the Norseman as my favorite airplane.

My first DC-3 landing at Lynn Lake would have been nearly perfect if I had understood that one must pin the DC-3 down upon initial contact with the runway. We approached the Lynn Lake strip and my wheels gently rolled onto the surface; the next thing I knew we were thirty feet in the air again. The oleo legs (like shock-absorbers in a car) compressed as the roughly 25,000 pounds of airplane settled on them, but the springs recoiled.

Because we still had flying speed, CF-CUE was forced back into the air.

Bob Stanley recommended that I come forward with the control column to "pin her down." My next approach was much better. While we didn't touch quite so gently, I was able to pin the airplane down the instant the wheels touched. Future landings were much smoother.

Sometimes of course, we would have a few fishermen on the Lynn Lake flight and Bob Stanley was not captain. On those occasions I remained at Flin Flon, providing weather reports and communication for the flight. I will always be grateful to Bob Stanley for allowing me to gain experience with DC-3s.

Finally, the Department of Transport reinstated me to full healthy pilot status. On June 2, 1958, I flew Cessna 180 CF-IEF to Lynn Lake with chief pilot Doug Rose as a pilot check-ride. When we arrived at Lynn Lake, Rose gave me a check-ride in Norseman CF-GTP, and I was fully restored to active flying duty.

Not long after my yearly check-ride with Doug Rose, I had to fly to a lake north of Lac La Ronge. Norseman CF-GTP was an aircraft with an ammeter that showed only charge, with no discharge arc on its scale. During the flight I was unaware that the battery was going flat. When I arrived at my destination, I tried a call on HF radio but discovered I had radio trouble, or so I thought. I usually repaired radios at any base from which I operated. There was no point in removing the radio out here in the wilderness, with no equipment to test it. When I was ready to take off, I discovered my dead battery. I was nearly a hundred miles from the nearest help, but without a radio I could not request assistance.

I'd heard of pilots who had been able to start Norseman engines by hand, although I had never attempted it myself. I decided to give it a whirl and attempt to hand crank a Pratt & Whitney R-1340, 600-horsepower engine. I would have ignition, of course, because aircraft engines use a dual magneto ignition system. The electrical system was only for lights, radio and the starter.

I stood on the starboard float behind the propeller and pulled that nine-foot arc of propeller time and time again. I was able to turn the engine and she would fire, just before I reached top-dead-center. The prop would kick back counterclockwise two or three revolutions. My fingertips ached from the effort. I stopped and tried to think of an alternate plan for getting out of that northern Saskatchewan lake.

I had spotted a tent just to the west from my location, along the northern shoreline; I wondered if there might be someone who could lend me some assistance. I discovered the camp to be deserted, but there was a low-powered HF radio there. These radios were built by the Department of Natural Resources and were battery-powered, running about two watts of output.

I tried the radio and could hear weak signals. I called La Ronge several times without response; I was finally rewarded with a call. It was Brian Magill. He had been a student with me at the Winnipeg Flying Club and was a good friend. Although pilots are often competitors, we would always go to great lengths help a downed brother. Brian said he was on a flight to La Ronge and was now heading back to Flin Flon; he would deviate from course and drop in to help me.

Brian taxied up to shore a few yards from where GTP was parked, and we went to the camp where I had used the radio. There we found a canoe, which we paddled to near our two Norseman aircraft. Brian removed the battery from his Norseman and we installed it in GTP after we had turned her away from shore. Brian tied the canoe painter to my port float cleat and I started GTP, taxiing out onto the lake. At that point he removed the battery from GTP and paddled it back to shore in the canoe. He returned the canoe and reinstalled his own battery. He had saved the day. In a couple of minutes, we were airborne and flying back to Flin Flon in loose formation.

10 Aimie Lake Excitement

On the morning of June 22, 1958, Oscar Erickson flew two Jewish brothers from Chicago into Aimie Lake for a full day of fishing. Both men were in their forties.

Aimie Lake had at least two positive things going for it: pickerel (walleyed pike) fishing was outstanding, and it was only twelve miles by air from Flin Flon. For anyone unable to afford a flight into the lake, a dirt road would take them within six miles of their destination and they could then walk the remaining distance. Local fishermen had left a boat on the lake where they had built a dock inside a small bay at the south end. They kindly allowed visiting sport fishermen to use the boat.

Oscar was to pick up his American clients in the evening, but he had been out flying all day, and when he arrived back at our base, he was tired and hungry. He told me that he just had to go home for his dinner before going to pick the men up from Aimie Lake. My day had been a light one, so I told Oscar not to worry about the men; I was only too happy to pick them up.

Oscar had been flying Cessna 180 CF-IEL all day, and I had been flying Norseman CF-GTP. It would make a nice change to fly Cessna IEF since it was much lighter than the Norseman and had such good performance.

As I approached the little bay from the north, the men were waiting on the dock; I entered about level with the tops of the hills that formed the narrows between the bay and the main body of Aimie Lake. Just as I was midway through those narrows, there was a down draft caused by the west wind blowing over the hill to the west of the narrows.

I made a mental note that upon takeoff I should stay well below the hills and thus avoid the down draft.

We loaded their catch and gear on board CF-IEF and started our takeoff run. We became airborne before reaching the narrows and I decided to hold the Cessna down near the water. Our speed was up to about 105 miles per hour as we came through the narrows, when I suddenly felt what seemed a heavy object fall on our tail. Our Cessna 180 was suddenly climbing at a very steep angle even though I had not moved the control column. I raised our flaps entirely from that I had used for takeoff and slammed the control column forward. CF-IEF lowered her nose toward the horizon, but not to the point where I could say we were in level flight. We were mushing along in a nose-high attitude at about 200 feet due to the very abrupt climb.

Without warning, IEF's nose went down and we found that we were looking straight into the lake. I hauled the control column back to my chest and applied full flaps, as I increased engine power to full takeoff. Now IEF's nose came up but it would not come up to level flight attitude. We were flying level but with a nose-down attitude, much like a de Havilland Otter does when climbing out with flaps extended.

Again, without warning, my Cessna went into that same steep climb and again I went through the motions of reducing flap, lowering my engine output and keeping the control column fully forward. There seemed little I could do, beyond making an effort to maintain some semblance of control over this airplane gone wild.

The steep dive and my tactics to control it were repeated several times. I had no way of keeping track of how many wild gyrations we experienced, as can be appreciated. IEF's nose went down for a final time without any indication that it would not respond to my tactics. However, I must admit the airplane did feel slightly different.

Tentatively, I reduced the throttle setting, but with full flaps and the control column nearly all the way back into my chest. We were slowly losing altitude, and I hoped that we might make a safe landing on the lake. I was concerned that the nose was too low and there was a good possibility that our floats could dig in and flip us on our back.

I experimented with throttle settings and found that I could control our descent by increasing or reducing engine power. Finally, we were at the point of touching down. The bows of our floats touched and tried to dig in. At that moment, I opened to full takeoff power; IEF danced on her float bows and I was sure she would flip onto her back. Our guardian angel must have been with us. Cessna 180 CF-IEF danced along on the tips of her float bows. Due to the heavy drag of the water, we were relieved that as she slowed, she finally seemed ready to settle down. 1 pulled the throttle back and the floats rocked back to level.

During those gyrations, nobody spoke a word. I'm sure everyone was occupied with their own thoughts, because we all were convinced they might very well be our last thoughts.

We had gone about two miles from liftoff to touchdown. As I taxied IEF back to the little bay and our dock, I tried to reassure my passengers that the danger was past. The two Jewish gentlemen were much more concerned that news of our near-disaster might reach the ears of their father, who might never allow them to return to Canada for a fishing trip. They explained how difficult it had been to convince him to allow them to come on this trip. Considering both brothers were in their forties, their father was obviously a force to be reckoned with.

At the dock, I examined the tail of IEF to discover what had caused our flight problem. Many aircraft make use of a trim tab on the elevator. Trimming is necessary to compensate for different positions of the center of gravity (C-G) and varied power and speed settings. A trim tab is a small area on the elevator that uses air flow past the surface. It aids the pilot's efforts to keep the elevator at the proper angle, without undue need for the pilot to constantly keep back or forward pressure on the control column while maintaining level flight.

The Cessna 180 uses a different system for trim. The horizontal stabilizer is fitted with two screw jacks that raise or lower the leading edge of the stabilizer. The screw jacks are controlled by cables which pass around pulleys on the trim control and screw jacks.

These jacks have only a single set of threads (one is molded in the movable part of the stabilizer and a bolt is inserted to act as the remaining thread pair). The hole through which the screw passes is slightly larger than the screw and, once in place, the bolt is inserted. It is a good system—provided the bolts are in place.

The hour was late, but I radioed for help. Gordon King was expecting my return and would be at the base, but the bookkeeper would probably have turned off our base radio when she left. As I suspected, I couldn't raise Gordon at our office. After a number of unsuccessful calls, I heard from Jock Cameron, who was base manager at the Lynn Lake base. He could hear me calling but was unable to copy voice messages. Jock asked if I could send my message using the microphone button as a radio-telegraph key. In addition to being a pilot and base manager, Jock was also an amateur radio operator;

copying Morse code was second nature to him, as it was to me. There are a few radio hams today who wish to do away with Morse code as a requirement for holding a ham license, but I feel strongly that this would be a mistake. We proved that day that Morse code can be very useful to pilots, as well as ham radio operators.

Once Jock Cameron had received my message, he telephoned Oscar at his home in Flin Flon to tell him about our problem. He asked that Oscar get Gordon King and fly him out to us in our Norseman CF-GTP. When they arrived, darkness was beginning to fall. Gordon King went to the tail of CF-IEF and removed the tail cone (called a bee stinger because of its shape). The bee stinger is aerodynamically streamlined as well giving the tail of the Cessna 180 a distinctive appearance. Gordon King straightened up, tears of remorse streaming down his face.

Some weeks earlier, Gordon had been doing 100-hour check inspection of the Cessna trim system. He had removed both bolts to check the assembly for wear. During reassembly, he had placed the elastic nuts on the bolts, but only finger tight, when the telephone rang. When he returned from the phone, he noticed that the elastic nuts were in place. They hadn't been tightened, but King thought the job was complete and he reinstalled the tail cone without making absolutely certain that the bolts were secured. No trace of either elastic nut was found; only a single bolt could be found, lying at the tip of the bee stinger tail cone. Gordon King's error had very nearly cost us our lives.

Inside the slit in which the stabilizer moves, there is a slip fence. The purpose of this fence is to seal off the opening into the tail from outside air and to keep water from splashing into the workings of the trim mechanism. A stop was built into the assembly, to prevent the stabilizer from moving too far up or down. During our wild ride, the upper stop had broken and that slip fence, on one side, had popped out of the slit, locking us in the nose-heavy position. However, I was glad it had happened that way. There could have been a much worse outcome if the stabilizer had not been held in a stable position.

Darkness was falling when Oscar took me aside and asked that I fly Norseman GTP back to Flin Flon. (I knew that recently Oscar had become nervous about flying after he ran a fuel tank empty and check pilot Rose had to override Oscar to switch to a full tank.) I was only too happy to take GTP back to Schist Lake. I also thought it would be good therapy for me to climb right back into the pilot's seat after my near mishap.

Two weeks after our near disaster at Aimie Lake, Oscar came to me with news that he had been offered a chance to reenter his old profession as an assayer. Oscar was my senior by nearly twenty years, but I was the base manager.

I was sorry to see Oscar go, but under the circumstances I felt it would be a wise decision. He was a very generous man, always willing to share, with one exception: he never offered a sip of coffee from the thermos that he always carried in his airplane. One day I found out why, when I removed the lid and smelled a heavy aroma of alcohol mixed with his coffee. Poor Oscar was only able to find courage to fly in a bottle. He was one of the finest men I have known and had proved his mettle as one of Canada's fine bush pilots. He had simply not been able to overcome his fear.

Bush and Arctic Pilot

11 Wing-strut Bolt Check

Following the experience at Aimie Lake and Oscar Erickson's departure, I did my best to encourage local prospectors and others to use our services. Slowly and reluctantly at first, new clients tentatively decided to give me a chance; one trip at first, and if we did well on that one perhaps there would be more trips. It was slow at the beginning, but gradually we began to build our business. It was my goal in the future that we might gain 50 percent of the flying out of Schist Lake. I felt that Parsons should receive about half of the local business. I'd be happy for the present, however, with 25 percent of the business.

Since I had flown with Parsons previously, I was familiar with his high-quality service and applied his methods to my own customers. I kept in mind that clients could drop me like a hot potato and return to Parsons. I began to enjoy my job as flying base manager; I was getting a good deal of flying under my belt and yet was able to keep my hands on the day-to-day operations.

Young George Dram had now joined my staff. He was an excellent pilot. I had every confidence that he would service our customers to the high level of satisfaction that I was in the habit of doing. He never let me down and in fact exceeded my expectations.

He was born at Cross Lake, Manitoba, in 1940 to his Yugoslavian-born dad and Métis mother. George was a mixture of English, Cree and Yugoslav bloodlines. Originally, the family name had been Drazenovic, but at some point the name had been changed to Dram.

At some point young George moved to Wabowden, where he learned to fly in a TransAir Mark VI Norseman aircraft. George was a natural pilot who learned to fly on fish pickups long before he had his license. He had built more than 100 flying hours in the Mark VI Norseman before he took flying training in Winnipeg. The instructor found this young man exceptionally gifted, considering he had never been to flight school. After George acquired his commercial license, he was hired by TransAir Ltd.; he was about eighteen when he joined my staff as line pilot at our Flin Flon base on Schist Lake.

George and I never exchanged a cross word with each other. He made my job much simpler because he just checked the book and did whatever flying had been assigned to his airplane, without any need for me to say a word to him. It seemed incongruous that George should be stuck flying our Cessna 180 CF-IEF when he had more hours in a Norseman than I had at that time. I had flown many types of aircraft and I'd flown many more than the required 1,000 hours. But to be insured on a Norseman, he needed 1,000 hours. I'd have trusted George with Norseman CF-GTP any day. Growing up in the bush of northern Manitoba had given him a natural sense of direction; he was a fine navigator.

On the morning of September 15, 1958, Norseman CF-BSJ was approaching to land near Nakina, in Ontario. A wing strut bolt fractured and the pilot was killed. With unprecedented speed, Ottawa contacted all Canadian Norseman operators by telegram, telling them to ground their aircraft until those wing-strut bolts had been checked.

There are two tests to check for possible bolt cracks on each wing. During a magnetic test, the bolt is magnetized and then covered with very fine iron filings. Any crack will show up, because the iron filings will stand on end at the edge of the crack. During a dye check the bolt is immersed in a mild acid bath and then brushed with a special dye. When examined, any cracks will show up as very fine bright purple lines, which means the bolt has a crack in it. Even a crack so fine that it cannot be seen by the naked eye can be detected by this method. If no cracks are found, the bolt is washed and dried and reinstalled.

I was on a fish haul when the telegram arrived. When I arrived back at Flin Flon, Parsons Airways Northern Ltd. had their Norseman tied to the dock; they had a good hour headstart on me. I went into town to hire some men to help remove the bolts. Hank Parsons had already been to town and he had hired every able-bodied man. He was determined to beat me on this wing-strut bolt issue. Since I could not do the job in a conventional way, I decided to sit down and think; there had to be a way. Suddenly it became crystal clear.

We had a large stepladder-like affair used as a refueling stand during winter. Gordon King, George Dram and I moved the stand to our dock under the left wing-tip of GTP. Eight inches of clearance existed between the top of the stand and the wing tip. Several blankets were folded and used as a cushion between the two surfaces. We then placed empty fish tubs on the port float and began filling them full of water. These were standard wash tubs that could hold 100 pounds of fish.

As ballast was added, the left float settled lower in the water, bringing the wing tip into contact with the stand. Gordon stood at the strut position holding a punch; when he sensed the bolt pressure holding was reduced, he signaled for us to stop pouring water into the tubs. King then drove the bolt out, leaving the punch in place while he did a dye check.

In probably less time than it takes to write about it, King had the original bolts checked and reinstalled in the left wing. The ballast water was dumped and GTP turned around for a repeat of the bolt check and reinstall on the starboard wing. All four bolts checked out sound, and in only a few minutes I was airborne again.

Hank Parsons was furious. He came storming over, demanding that our aircraft be grounded. Gordon King assured him that GTP's wing bolts had been checked and were in perfect condition. Parsons wanted to know how we had done it so fast (by now our refuel stand had been stored away). Gordon told him that I had requested that we not give away our secret until after Parsons Airways had completed checking their own wing bolts. When I arrived back from my next flight, Parsons was waiting for me on our dock. "How did you guys check them bolts?" he asked. "I hired every available man, so you pot-lickers had no help."

"When the opposition hires all the available manpower," I told him, "it becomes necessary to go for brain power."

When I showed him how we had accomplished the task with much less muscle power than he had used, Parsons swept his hand to his ever-present baseball cap, grasping the peak with his thumb and forefinger. He pushed his cap onto the back of his head and then scratched with his little finger. "Jeez," he said. "Why didn't I think of that?" At this point we both had a good laugh. Even though we were competitors, I was comfortable knowing that should I ever come to grief, Hank Parsons would be the first to come to my aid. Hank also knew that I would be there if he was the one who ever needed help.

12 Joe Bear and His Boy

Returning from a charter to the south shore of Highrock Lake, north of the Burntwood River during the summer of 1957 in Norseman CF-GTP, I flew directly over Sherridon. There was a man on the dock waving a white towel. Sherridon was no longer active with mining operations, so air travel to it was not an everyday occurrence. I decided this might be an emergency; I landed and came alongside the dock.

"I'm Joe Bear," the Cree man on the dock said. "This is my little boy. Three or four days ago, he cut his finger, but now he's got poison. I waved you down to see what you think, anyways."

"You're right, Joe," I told him, "I think that it's blood poisoning and we have to get your boy to the Flin Flon hospital. They'll know what needs to be done."

The man shifted uncomfortably, then said, "I can get him there by canoe."

"Joe, I don't think there is time. Climb aboard. I can get you there in just a few minutes," I said.

"I don't have money for an airplane," he replied.

"Don't worry about the money just now, we have to get your son to the doctor; we can talk about money later." Joe Bear and his son climbed on board and we flew directly to Flin Flon. I issued him a twenty-two-dollar ticket, which was the old fare when a scheduled service had operated into Sherridon.

Since the charter had already been paid from Flin Flon to High Rock Lake and return, it would have been better if I had not bothered issuing any ticket to Joe Bear.

Since our flying was finished for the day, I drove Joe Bear and his son up to the hospital in my 1955 Ford Customline sedan. Joe said he had friends with whom he could stay and that when his son was ready to travel they would find their way home to Sherridon. I imagined they would travel by canoe via some circuitous route about which I knew nothing. The ticket went to head office and in due course I received a terse note wanting an explanation about this twenty-two-dollar bill, which had not been paid. I wrote an explanation about the boy's need for medical help and the fact that this father was short of cash at the time. I had extended him credit, I said. In addition, the fact that I was on a charter that had been paid meant that we had suffered no loss even if Joe Bear's ticket was never paid.

A few weeks after my explanation, I received another note from the same head office man who reminded me that company policy (actually, I'd never read it) stipulated that should anyone be found, sick or injured and unable to pay for transport, they were to be left where they were. A far cry from my old boss Hank Parsons, who had a policy that if anyone was found out in the bush either sick or injured they were to be picked up without regard to their ability to pay.

I replied to TransAir that I would always transport anyone who was in need without considering their ability to pay. I felt this would answer their concern, but I was wrong again. About six months later I received another head office letter. "A good policy in considering this type of credit extension would be to consider it as being your own money." I was angry. I scribbled "over" on the face of that letter and wrote the following note: "With regard to the Joe Bear account you suggest that I consider this to have been my own money. To this end I hereby request that the amount of twenty-two dollars covering this ticket be deducted from my next paycheck. ... Company policy in this regard might be carried out, but only by another base manager. As long as I hold this position, my policy, which happens also to be our competitor's policy, will prevail."

That did it. No further letters or notes were received and no deductions were ever made from my paycheck.

Almost a year following the pick-up of Joe Bear and his son, I found myself walking along a Flin Flon street when I was suddenly called from somewhere in the crowd of people.

"Hey, pilot!"

I turned to see who was calling. "Hey, pilot, wait up," he shouted and then I saw a Cree man waving his hand. I waited to allow the man to catch up with me. "Hey, pilot," he said again, "I owe you some money."

I looked at him and replied. "No, I don't think so—"

"Don't you remember? You picked me and my little boy up. He had poison in his finger, eh?"

"Joe Bear!" I said, "I had nearly forgotten you, sorry. How are you and how is your son?"

"Oh, he's good now, no more poison, eh?" he said. Then he pulled out his wallet and gave me a twenty-dollar bill, together with a five. "Here, I always pay my bills." I reached into my pocket to hand him his three dollars in change. I didn't want to overcharge him, but he refused it.

"Listen, pilot. I appreciate what you done for my little boy, eh? Go buy yourself a beer." We shook hands, and suddenly Joe Bear was lost in the crowd.

The twenty-two-dollar submission was prepared and mailed with a covering note telling my head office contact that in order to be a good corporate citizen we must sometimes trust our fellow man.

Bush and Arctic Pilot

A. R. (Al) Williams

13 Arctic Adventures

Overseeing TransAir's Flin Flon base and making the business grow was interesting and a challenge. I decided to visit a number of the fish camps in the area and made contracts with the fishermen so I could haul their fish to market with Norseman CF-GTP.

My agreement with them was that I would haul fish for them at seventy-five cents per mile for 1,800 pounds of fish.

In the past they had been promised this deal (seventy-five cents per mile was our regular charter rate), but if a shortage of Norseman aircraft would occur and a substitute airplane was sent in, the fishermen were charged the going charter for that substitute aircraft. A Cessna 180, for example, would carry around 600 to 800 pounds per load, but since the charter rate was forty-five cents a mile, if only 600 pounds were carried, three trips for each 1,800 pounds were made, bringing the total charge for 1,800 pounds to a dollar thirty-five per mile.

My contract stated we would move the fish at the seventy-five per mile rate without regard to what aircraft we used. Our Norseman CF-GTP was being put to good use. I was happy, because I was getting lots of flying and the company was making money—but not for long.

Stan Wagner didn't seem to like my base doing well, and he called me in August 1958 with an order that I take my Norseman to Churchill.

On the evening radio sked, I spoke with a pilot at Lac du Bonnet who was unhappy because there was little flying for him and his Norseman.

I telephoned Wagner to protest my going to Churchill when I had fish haul commitments, particularly when a Norseman was sitting at Lac du Bonnet with little or nothing to do. Why couldn't we, I asked, have that Norseman go to Churchill? Wagner replied, "Well, fella, you will just have to do what you're told to do."

"But this is poor business, Stan," I told him. "I'm going to be away from these fish contracts while a Norseman at Lac du Bonnet is essentially sitting idle."

"Well, fella," Wagner said, "Are you refusing to go to Churchill?"

"No, I'm not, Stan. It's like this, Al Williams the pilot wants the experience of going to Churchill. Al Williams the base manager has contracts which must be looked after. If I am in Churchill, George Dram is going to have to haul all the fish with the Cessna 180."

In the end, I had to go to Churchill, but I wrote a letter for head office files protesting the move on the grounds of inefficient use of aircraft, as I had fish contracts that needed servicing by Norseman GTP.

Upon my arrival at Churchill a day or two later, I was told that Norseman CF-GTP had to be equipped for instrument flying. Almost at once my seaplane was put onto a flatbed trailer and moved to the hangar two or three miles north of Landing Lake. Installation of "plumbing" to which various vacuum-operated flight instruments would connect was made. This instrument flight package was necessary for my survival while flying in bad weather. Bad weather was not just a possibility; it was a certainty, because Churchill is notorious for its foul weather. In fact, I believe Churchill's main industry is the manufacturing of foul weather, which it exports to the rest of the world.

I was pleased to meet Charlie Webber, Sam Bonderoff and several other seasoned arctic pilots in Churchill. Jan Malinski and his copilot Chuck Fenwick were flying the Canso, doing ice patrols.

As I waited for GTP to be equipped for instrument flight, I borrowed a company pickup and went to the Churchill town site. I found myself on the shore of Hudson Bay, across the bay from Fort Prince of Wales, watching seagulls and enjoying the sounds and smells of the waterfront.

The fort was constructed by the Hudson's Bay Company around 1733 and eventually was destroyed by the French about 1782. As I started around a large boulder, I stopped in my tracks.

Less than a hundred feet away, I spotted a very large polar bear. I was thankful I was downwind from this powerful animal; I was able to retrace my steps without the bear catching my scent. These beautiful and deadly carnivores are capable of moving swiftly. It seemed unfair that such beautiful animals, which seem so cuddly, are so lethal.

Stan Wagner arrived in Churchill on board our DC-6 service from Winnipeg to Montreal, with stops at Churchill and other settlements along the way. Wagner was in Churchill on a routine inspection, or that was his story. He entered the hangar and demanded to be told why Norseman CF-GTP was not out flying. The engineer making the installation, Dave

Johannesson, told Stan that I would not have my airplane until it was fully equipped for instrument flight. The exchange between the two men was related to me by Johannesson.

Wagner: "I want this airplane flying out of Landing Lake right away!"

Johannesson: "This airplane is not leaving the hangar before we have installed the plumbing and instruments."

Wagner: "I just gave you an order. You will follow that order or you will be walking down the road."

Johannesson: "None of our aircraft fly out of here without instruments. This Norseman is not going to be the exception." Wagner: "You'll do what you're told, or you can pack your bags and leave."

Johannesson: "Fine. Fire me and I'll write up a snag on every airplane on the base. Each one I write up will take a lot of disassembly to find. You won't turn a propeller for a month."

Wagner backed down on the issue of instruments for my Norseman. There must have been a reason behind his insistence that I fly without instruments. I leave it to the reader to determine. One thing I can say without hesitation: If I had flown Norseman GTP out of Churchill without flight instruments, I would not have survived. Upon my return to the hangar I found GTP would be ready to fly in the morning. Our dispatcher Gene Selinski had already booked a trip out of Churchill to Bob Kennedy, Indian Affairs and Northern Development Agent for the area. An Inuk had died nearly a year earlier. Kennedy was to investigate and write a report on the man's death. When I left Gene Selinski and walked outside into what had been a beautiful and sunny day, I stepped into an arctic front with fog so thick I had difficulty finding our ex-military "H" building. The weather had changed in less than thirty minutes. Now 1 knew why an instrument package was a necessity for all aircraft flying out of Churchill.

The first leg of my flight with Bob Kennedy was Churchill to Eskimo Point (now Arviat, Nunavut), 170 miles along the west coast of Hudson Bay. We were to pick up a Royal Canadian Mounted Police special constable who was to fly with us to where the Inuit man's death had taken place. The RCMP had a number of Indigenous "special constables" throughout the north for such occasions. These constables were not full RCMP officers; they acted as liaison officers and originally were hired as scouts because of their knowledge of the country. Before leaving Churchill, five different pilots and other staffers told me about Padlei, north of Churchill. Should I need fuel, it would be at Padlei; if sleep was needed, I should go to Padlei. I was naturally convinced that Padlei must be an important, well-equipped town site, even though I had never heard of it.

Shortly after takeoff from Landing Lake, we crossed the shoreline beside Churchill's rather shabby town site. It seemed a bit incongruous that the grain-handling facility was modern and well-kept, in contrast to the drab houses of the town.

Bob pointed out the cockpit windows of GTP and shouted for me to look down at all the white whales. I was confused at first because I saw nothing that was white beneath us. What I did see were a large number of green shapes directly below us; in a moment I realized we were seeing a pod of white whales swimming in the green-tinted waters of Hudson Bay.

The weather was cool but fine, and I began to question the need for the flight-instrument package that GTP now carried. About forty miles up the west coast of Hudson Bay, we crossed the northwest leg of the Churchill radio range. That information was filed for future use, although the range leg was also printed on the map. The balance of the flight to Eskimo Point (Arviat) went smoothly, and we arrived just in time for Sergeant Bill Gallagher to invite us for lunch. Gallagher introduced us to an officer named Lou and to Special Constable Jimmy Gibbons.

Gibbons is not an Inuit name; when Jimmy was only a few months old, his parents had died and he had been adopted by Constable Gibbons.

After the introductions, I went to the Hudson's Bay Company (HBC) and found Jimmy Gibbons sorting mail. The HBC office was cool by my standards, about 60 degrees Fahrenheit; I found Jimmy sorting outgoing mail into three bags for three destinations. Sweat was running from his forehead, while I shivered in my parka. He was accustomed to much cooler weather than this. When he had completed his tasks, we had lunch, and after refueling, we departed from the bay on the north side of the Point and settled on a heading that would carry us to Padlei—even on my first day in the Barrens, I was to visit the famous Padlei.

The Churchill hangar crew had installed a vertical speed indicator, directional gyro and artificial horizon in the panel of my Norseman. I felt I should become familiar once again with the art of flying an airplane by watching the quivering needles on instrument dials. I had learned to fly instruments in Winnipeg, but it had been several years since I had actually flown under real instrument conditions. Because Bob Kennedy was familiar with the country over which we now flew, I asked him to act as a lookout so I could concentrate on honing my rusty skills.

For the first few minutes, I experienced some trouble in concentrating on maintaining directional stability while continuing a steady climb to 5,000 feet. I chose this level because most of the flying in this area was below 5,000; this would reduce the chance of conflict with other air traffic. When we reached our cruising altitude, I was able to settle into a routine of scanning instruments; my old skills started to return, albeit slowly. I must have maintained a relatively good track, because in due time Bob shouted, "Not bad! There's Padlei, right on the nose."

Search as I might, I was unable to see a town site. Then, right in front of us, I made out three buildings, with firewood stacked in the shape of tepees so the strong winds of winter wouldn't bury the wood under snowdrifts.

We landed in the river at Padlei, picked up HBC manager Henry Voisey and departed for Yathkyed Lake, some fifty miles northwest, to search for a nomadic Inuit campsite. Because of the fading light—it was 22:00 hours—we were unable to locate the camp, and we returned to Padlei for the night. That night, Lou and Jimmy took their fishing tackle in hopes of catching some Arctic char. Half an hour later, we heard someone laughing as though his sides would split. We ran outside to find out what was happening. Lou had stepped on an algae-covered rock and had slipped into the river. Jimmy was laughing so hard he was scarcely able to help Lou out of the chilly water.

Next morning we flew to Yathkyed again; the weather was ideal so we had no trouble locating the area. I flew directly across the lake at 1,000 feet as I checked for landing hazards such as rocks, reefs and sandbars; I noted rocks near the shoreline all around the lake, but none beyond fifty yards out. Timing my flight across the lake, I learned there was plenty of landing room, so I came around onto final approach. The landing approach was quite normal although rather shallow; I didn't worry about that. I just put it down to my inexperience with this magnificent and beautiful land.

Three tents marked the camp off to our right as we settled toward the lake; the lake seemed much smaller than I had originally thought. I reduced power and as soon as the keels of my floats touched the water, I hauled back on the control column to dig in the heels for a quick stop. When we had nearly stopped, it became clear we had quite a long way to taxi in order to reach the first of the rocks on the south shore. This seemed rather strange, until I realized we had landed on a lake at the bottom of a shallow bowl. It was about three miles in diameter and deep only in its center, where it was about 300 feet deep. This explained why our approach had seemed so shallow; we'd been traveling eighty miles per hour, descending 300 feet per minute, so we had remained nearly the same altitude above ground all the way in to landing. I decided we would need to climb at about 350 feet per minute upon our departure in order to clear the rim with room to spare when leaving.

We all watched for rocks near the south shore, and when I positioned GTP just outside the rocks, I stopped my Pratt & Whitney wasp engine. The passengers stepped into waist-deep, frigid water and began to wade ashore. I had to remain with the aircraft to keep her from drifting away, so I taxied away from shore and dropped my anchor with 150 feet of line and allowed GTP to drift until the anchor snagged bottom. It was strange to witness the men walking toward that campsite, since distance was impossible to estimate. The men didn't seem to be making headway; they only appeared to become smaller. This was my first experience of a strange phenomenon that seemed to me to be unique to the Arctic; perhaps it is also experienced in deserts.

As I watched over my anchored Norseman, the men stopped for only a minute and then retraced their steps back to the lake. I began taking in the anchor rope, and as I pulled, my Norseman proceeded toward the spot where the anchor lay beneath the surface. I had to pull GTP directly above that spot before I could weigh anchor. The anchor had snagged on some large rocks on the bottom and needed a vertical pull to dislodge it from those rocks.

By the time I had taxied back near shore, the men, shivering in their wet clothes, had waded out to my position and climbed onto the starboard float. Poor Jimmy however, slipped on a submerged rock and flailed his arms trying to stay vertical. He failed, and as he went down, only his fingertips reached the starboard float. We all had a good laugh, this time at Jimmy's expense.

It turned out the camp was deserted, but signs of a large herd of caribou heading south gave us clues that led us to believe the Inuit had followed. Takeoff was quite short, and I immediately established our rate of climb near 350 feet per minute. Again, it was a strange sensation to remain at nearly the same altitude, above the ground, for at least a mile after lifting off.

We had gone perhaps thirty miles south of the camp when Kennedy shouted to me over the roar of the engine. "Look at all the caribou!" I glanced down but saw nothing beyond the light brown moss and lichens with which the country was covered. Suddenly, I realized the big carpet of brown was moving. I had never before witnessed such a gigantic herd. The landscape was covered for many miles in every direction by caribou.

Perhaps two miles west and five miles beyond the herd we saw tents perched on the southeast end of another lake. It was there that we expected Bob would find his answers. Once more, I dragged the lake, seeking submerged hazards. Although some stones appeared fairly near the surface, a low pass proved I had nothing to fear from them because the water was very clear and rocks that seemed close to the surface would prove to be perhaps eight or ten feet under water. The campsite was beside a rapidly flowing river, which was only about four feet across and about a foot deep. This ideal location provided the people with fresh, cool water for drinking and cooking.

We landed from the west against an increasing east wind that blew directly offshore away from the camp. We had taxied slowly into the little cove where the camp stood when I suddenly saw we were surrounded by a field of rocks just below the surface. I pulled the mixture control and stopped the engine, cursing my foolishness at not being more observant as we entered the little bay. We were several hundred miles away from any assistance should we hole a float; I was amazed that we hadn't hit one of these stones, because we were well inside a ring of them that guarded the entrance and partially filled the harbor itself. We had little chance of escaping damage, but Bob and Jimmy went down on the floats with Lou and they began to paddle in reverse in an attempt to guide us around those dangerous rocks.

In the meantime, the breeze had increased, and we began moving astern, much too fast. I fired up GTP again, and with the throttle pulled back to slow idle we were able to check our drift. A few times, Bob called for more power to prevent our backing over a rock before he was able to position us to left or right. Miraculously, we extricated ourselves from the rocky hazard. We lost thirty minutes in doing so, although it seemed no more than five. Only once did we actually collide with a fairly large rock; we were able to slither over it with the keel of the starboard float lightly in contact.

I was able to beach the floats on a bed of fine gravel just left of the camp, across the stream. I turned GTP around and secured ropes from the rear cleats to a pair of stout rocks that stood about fifty feet back from the shore, so the heels of our floats rested gently on the fine gravel beside that stream. Those rocks were the only anchor points available, as the northern landscape offered no trees or shrubs. In a short time, the wind increased; I found myself fascinated by the tufts of moss and lichens ruffled by the wind.

The outside temperature had dropped and the increasing wind carried a good deal of moisture from Hudson Bay, making it bone-chilling and bitter. While Bob and Jimmy questioned people in the camp about the Inuk's death, I accepted a cup of tea from an Inuit woman. Jimmy would interpret from Inuktitut into English, and Bob wrote up his investigation on the man's death. My job was to provide transportation, and as such I was not privy to police details, nor was I curious enough to ask—I probably would have been told that this was confidential information. When all the questions had been asked and the answers duly noted in Bob's journal, we were ready for departure. Our flight to Eskimo Point (Arviat) was routine, since the weather held for the rest of the day.

I noticed that the caribou herd had moved only about a mile or two south of their previous position, which indicated to me good grazing in the area. We returned Jimmy and Lou to their home base and remained at the Hudson's Bay post overnight.

Next morning, I checked the weather picture for Churchill, which was clear with a temperature of 60 Fahrenheit; warmer than we had experienced on departure from Churchill. We left with three hours of fuel on board, which was more than sufficient for the slightly less than two-hour trip to Churchill.

The first hour of flight was without incident, beyond the fact of a ten-degree left crab angle to compensate for the east wind. We were encountering some clouds now, which were beginning to form at our altitude. I needed to descend to 1,500 feet to maintain visual contact with the shoreline.

By the time we crossed the northwest leg of the radio range, we were into much lower cloud and were forced down to 500 feet. My check of Churchill's weather gave little encouragement; they were reporting a 300-foot ceiling and a visibility of two miles. We had plenty of fuel for the rest of our flight, but we had passed the point of no return, as we could not return to Eskimo Point on the fuel we carried. I did not consider it an option to fly instruments, as I had not yet honed the needed skills. We were forced down to about 400 feet as I visually clung to the shoreline.

With visibility so reduced, I had difficulty understanding why my directional gyro readings seemed wrong. We appeared to have taken up a heading of 30 degrees, which must mean we had turned to the left some 150 degrees. Yet I was still in visual contact with the shore of Hudson Bay. It took me a while before I finally figured it out—we had been following the contour of Button Bay.

Because of the shape of Button Bay and our present heading, I reasoned that the mouth of the Churchill River must be close to hand. I queried Bob Kennedy to this effect, and he quickly confirmed my theory. He was familiar with this part of Hudson Bay from extensive trips by boat. He was not upset by our situation; in fact, he seemed to enjoy viewing the shore whipping by at well over 100 miles per hour from our vantage point of fifty feet. Ceiling and visibility had deteriorated to the point where we were forced to this altitude, and I was feeling uncomfortable; I was not inclined to land in that rough water, particularly when I knew the Churchill River lay such a short distance ahead.

Suddenly the shoreline took a very sharp turn to the right and was nearly lost from view as I made a tight turn in order to stay in sight of it. Bob told me, again from his storehouse of experience, about a very steep cliff well inside the mouth of the river, and suggested I land before reaching that cliff. I could not fault his logic, so I made a 180-degree turn to port and set up a landing approach with nothing but the waves to guide me. I could not see the shoreline to either port or starboard. I did not realize it at the time, and quite possibly my lack of experience would not have allowed me to evaluate it, but the wind was blowing directly into the mouth of Churchill River, while the tide was ebbing, a situation that makes for very rough water. We touched one wave and bounced and the entire world disappeared. The base of the clouds had dropped to about thirty feet.

Bob Kennedy now looked concerned; he had quite obviously been comfortable until that moment. I applied a bit more power as we settled back into visibility and we touched again, but this time we didn't bounce and old GTP settled onto the water. Now it was my turn to feel less concerned, but we still had to find our way to shore.

I called Churchill Radio, advising them we had landed somewhere in the mouth of the river but that I could not see shore, and asked for suggestions. I was told to hold my microphone button closed for a minute while the Harbour Board took some bearings on my signal; they would then give me a heading that would take us to the beach beside the grain elevator. As my microphone button was held down, the sound of a bell reached my ears from somewhere not far away. At the end of my minute, I mentioned the bell sound. I was instructed to kill my engine, report my heading and the relative bearing to the sound of the bell.

The tide was carrying GTP toward Hudson Bay, and I was told to start GTP and take a heading of 120 degrees because I was in danger of drifting into a buoy anchored in the river mouth. I had not yet gotten GTP's engine started when I saw the buoy looming out of the mist. My Norseman was rolling with the waves. The buoy, a big black cone approximately twenty-four inches in diameter at water level and tapered to six inches at its apex, was bobbing to and fro about 45 degrees to left and right and at nearly the same height as my wings.

We drifted toward that buoy and bumped gently into it on the outboard side of our port float. GTP began to drift, tail first, around the cone. I was mesmerized by the sight of our port wing as it rose to a steep angle while the buoy bobbed under it; I could visualize the wing coming back down and the buoy spearing it through, possibly breaking the main spar. Fortunately, the buoy bobbed the opposite way as the wing tilted to the left. It happened five or six times before we drifted out in the clear, and the only damage we suffered was a broken Pitot mast, which came down dead center on top of the buoy during our last roll.

Once clear of the dreaded buoy, I started the engine and took up the new heading dictated by the Harbour Board. I was probably less than 200 feet from shore when Bob and I saw the beach at the same time. I must say, we were happy that this particular mission had come to a safe conclusion.

The following day, after a new Pitot mast had been installed on my port wing, I was directed to Chesterfield Inlet (Igluligaarjuk), on the west shore of Hudson Bay. My mission was to gather up Inuit kids and transport them back to residential school, following the summer break. Only a few trips with the school kids had been completed when I was asked to fly to Arctic Bay on the north end of Baffin Island (at latitude 73 degrees 19 seconds north and longitude 85 degrees 11 minutes west) to support an American crew involved in preliminary work on an iron mining operation. I was to provide transport for some of the men and gear to Churchill, where they could return to the U.S. prior to freeze-up.

With full tanks, I took off from the small lake at Chesterfield Inlet and turned on a heading that would take me to Repulse Bay (now Naujaat). The flight to Repulse Bay, which straddles the Arctic Circle, was an adventure in its own right. As I over flew Devil's Gap, I knew it had been appropriately named; ragged black rocks seemed to have teeth that reached hungrily up toward my floats. For a moment, I experienced a touch of fear and foreboding.

I can't recall the names of all the people I met in those early days of the 1950s, and my log book records only my aircraft movements: hours, minutes and miles involved with the business at hand. I owe a debt of gratitude to the man at the HBO post at Repulse Bay, whom I will refer to as Bill Jensen. He became my radio lifeline in the days that followed. "Each morning between the hours of six and nine," Bill said, "I will call every hour, on the half hour, with weather information and for your position report." Bill looked at the shoreline and went on. "In the evening from six to eight, I'll do the same. I'm sure we'll make contact during at least some of those skeds."

The author at age nine beside Stinson SR-8 CF-BGW, Prince Albert, Saskatchewan, 1938.

The author at age nine standing by Waco Cabin biplane, on the North Saskatchewan River at Prince Albert in 1938.

Author on right fload of de Havilland Otter CF-ODK at Sious Lookout, Ontario, summer of 1954.

Cessna 180 CF-IXU during winter of 1956 at Flin Flon, Manitoba.

Office of Parsons Airways Nothern Ltd., Flin Flon, Manitoba.

Cessna Crane in RCAF colors; this was the type of aircraft being flown by Sgt. Jack Park when the left propeller struck the ice near Brandon, Manitoba.

CF-IRE burned at Eskimo Point, August, 1958. She was a dog!

Engine change in Norseman CF-DRD after blowing a cylinder heat at Hughes Lake, July, 1959.

The author (with necktie) as a Morse code radio operator.

Cessna 170 CF-H1Y.

CD-3 OF-CAR before forced landing—plenty of fuel here!

De Haviland Beaver CF-NOT in Edmonton.

De Havilland Otter CF-CZP at Cambridge Bay, N.W.T.

DC-3 CF-CAR on Tuhdra near Byron Bay on distant early warning line after running out of fuel.

This Douglas C-3 is equipped with tanks for carrying fuel to camps out of Red Lake, Ontarion, in 1992.

This Cessna 140 C-FXNN belonged to Bernie McIvor of Edmonton. This is exactly like CF-DMQ which the author flew out of Winnipeg Flying Club on the day the air force jet nearly ended his flying career! Note the spring steel undercarriage legs and stubs that extended the wheel axles forward, making the aircraft very "twitchy" on landing.

Red Lake Airways Norseman CF-GLI just arriving

Beech 18 CF-SRK at Cambridge Bay, N.W.T. This aircraft was damaged when it ran onto a soft spot.

DHC-2 de Havilland Beaver CF-OBS in Ontario Forest Service paint being made ready for a water-bombing attempt.

A Beech 18 taxiing out on Howey Bay. Many Beech 18s have found their way into bush flying in various areas of Canada.

Huey Carlson departs Howey Bay at Red Lake in 1993 in his Norseman CF-FQI.

DHC-3 de Havilland Otter tied to dock at Red Lake, Ontario, in 1992. Pilot Mike Richter stands on the dock.

Norseman C-FBHZ during a fly-past.

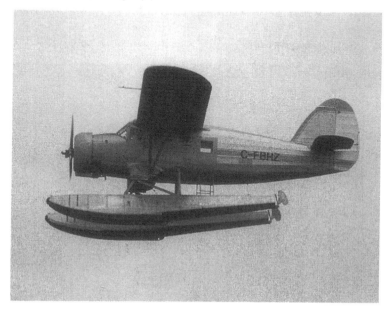

C-FJEC taxiing on Howey Bay at Red Lake in 1992.

Robert H. Noorduyn (left), son of the designer of the Noorduyn Norseman, with Glen Crandall, owner of the Norseman CF-UUD.

Norseman CF-DRD—last time on floats, early summer, 1992.

Norseman CF-DRD on pedestal in Red Lake. Front Streed in the background.

Mark VI Norseman C-FOBE at Red Lake—the only all-metal Norseman in the world, July 1993.

Cockpit layout of the Noorduyn Norseman. Home Sweet Home!

The author and a young Scorr McAllister beside C-FOBE, Red Lake, July, 1992.

Otter C-FMEL tied to the dock at Red Lake in 1992.

Norseman CF-BSB waiting to begin its fly-past at the Norseman Festival, July, 1992.

Another restored Norseman with a backdrop of the west shore of Howey Bay.

The author, looking like a plantation owner, with wife Louise at home in Edmonton, 1995.

Bill's outpost was equipped with the new Crosby Electronics SSB/CW equipment. "CW" means continuous wave and is a term used by radio operators for Morse code signals.

Because of limited electrical (usually battery) power, Hudson's Bay radio equipment was limited to ten watts. My Norseman CF-GTP was equipped with war-surplus command radio gear, also of limited power, using an inefficient screen grid modulation system.

I departed Repulse Bay in the late afternoon of August 13, 1958, and took up a heading for the settlement of Igloolik. The maps simply referred to the vast areas of land beneath the keels of my floats with the word "unexplored."

At times a lake would appear on my map, while no corresponding body of water could be seen from GTP. An hour later, perhaps, I would discover the mapped lake, removed as if by magic, to a new location 100 miles to the north. At other times a map landmark seemed to have moved miles to the left or right of track. In retrospect, I feel privileged to have been one of the northern pilots who experienced flying over Arctic terrain relatively unknown. At the time, however, I was less than comfortable.

The sun was low in the northwestern sky, descending at a very flat angle toward the northwest; I approached Hall Beach (now Sanirajak), Distant Early Warning Line (DEW) Site 30 in Foxe Basin, and was about to call Churchill on 5680 kHz with a position report. I anticipated weak signals and increased the audio gain of my receiver. As I reached for my microphone, I was surprised by a very loud voice signal.

"Churchill Radio, Nancy 3762 Charlie, d'you read?" the voice said. After thirty seconds I figured his gain would have been turned up, as mine had been.

"Sixty-two Charlie," I said, using phonetics of the day, "This is Charlie Frank George Thomas Peter, go ahead."

Several seconds went by and then the reply, "GTP, sixty-two Charlie, where y'all at?" I reported my position as near Foxe Basin. "Ahh ... that's impossible," he said, "we're just comin' up on Site 30; what type airplane are you?"

I told him I was a Mark VI Norseman, which didn't seem to help.

"What type are you?" I asked.

"We're a DC-3 from Dallas: out of Thule, bound for Churchill, what's your altitude?" I told him I was at 5,000 feet. He suddenly responded. "GTP, sixty-two Charlie, we've got you in sight; what're y'all doin' up here with one engine? We got two of 'em and that ain't enough." We both continued on our way.

I advised Bill Jensen, standing by for my position report, that I had arrived at Igloolik. The time was 23:00 hours and the sun was very near the northwestern horizon. Ice floes had drifted into the bay directly in my approach path. Those floes took on an eerie, otherworldly appearance under the backlighting influence of the setting sun. I hadn't expected to encounter ice floes and I was taken aback, quite unsure as to how to land GTP without hitting one or more of those jagged floating chunks. Some were as small as a bread box; others were larger than a multistory home.

I flew a couple of passes over the bay, which convinced me there was a way to land between the floes. Having landed, I began to taxi slowly toward the settlement, weaving to avoid making contact with even the smaller pieces. At one point I found myself in near panic; I saw what I took to be a very large rock in my path, just under the surface of that crystal clear water, only a few yards in front of my bows. I couldn't turn because of the ice; I pulled the mixture control to idle/cut off and scrambled to my left float; I grabbed the paddle from the clips on the inside of the float and began back paddling like mad, in an attempt to get the aircraft stopped before disaster. It turned out that I need not have worried; the water was so clear that what I had originally taken as a large rock was seaweed growing on the bottom of the bay perhaps thirty feet beneath the surface.

A good deal of activity was taking place on shore as I tied up to the buoy fifty yards offshore. Rising and falling tides made it necessary to moor at the buoy, rather than beach on shore. A long boat came alongside to transport me to shore after I had stopped the propeller in full coarse pitch (the salt air would otherwise attack the exposed barrel). In due course I discovered a whale had been killed by several hunters, who were dividing it so each family would receive their fair share.

Father Franche invited me to stay at the mission, since a survey crew had moved into the Hudson's Bay Company outpost. I had not eaten since morning and was glad to accept the good Father's offer; in place of the expected Arctic fare, he prepared a steaming plate of canned Burn's

Chuckwagon Dinner. I was surprised but pleased. The mission building was fairly cool (near 60 degrees). Father Franche explained that the supply ship had failed to reach Igloolik during the past year; while the vessel was now at Eskimo Point, no one could say with certainty that it would deliver a fresh supply of coal and other goods this year. No one wasted fuel by keeping warm during the summer when remaining fuel might be needed in the dead of winter.

At 03:30 hours I was awakened; a wind had sprung up and was beginning to move ice into the anchorage. I had to taxi GTP 300 yards east into a much more sheltered bay at once. During the next three days, the wind beat relentlessly from the south, driving those ice floes against the beach. It was impossible for my northward flight to proceed. Meanwhile, GTP was completely safe in her sheltered cove. The Hudson's Bay people contacted Bill Jensen and advised him I was a victim of wind and ice but otherwise all was well. The HBC men told me they would prepare a meal of muk-tuk and seal liver for the noon meal next day; I looked forward to having a taste of Arctic cuisine. Instead, the wind came about during the night, and by morning the bay was clear of ice. GTP and I were on our way toward Arctic Bay.

Traveling north over Admiralty Inlet, I came upon a fjord cutting 90 degrees across my flight path. This fjord was to lead me to the mouth of Arctic Bay. The scenery was magnificent. The entire area is magnificent. There is no vegetation evident, but rocks offer almost every hue of color imaginable. Those sheer rock walls, 1,000 feet high, plunge vertically into the fjord; the bay opens to reveal a large and beautiful harbor where the settlement of Arctic Bay is located at the base of mountain slopes, along the north shore. Those surrounding mountains run more than 3,000 feet high. Mountains in other parts of Canada can be much higher, but these Arctic mountains are at sea level.

I was able to contact Bill Jensen back at Repulse Bay by using the microphone button as a telegraph key. Although it was often possible to read his SSB voice signals, here he had to send CW to me as well. Thank God for Morse code. After taking on fuel at Arctic Bay, I flew east to the southwest shore of Ecstal Lake, about 1,200 feet above the Arctic Ocean. As a safety precaution, I dragged the lake looking for any reefs or submerged rocks. The clear waters of Arctic lakes came as a surprise. In the 1990s pollution may have muddied the waters, but then I was able to clearly see schools of fish swimming many feet beneath the surface of Ecstal Lake.

Even as midnight approached, I could not bring myself to sleep, as I longed to experience the midnight sun. The beautiful, cold, clear Ecstal Lake water, perfect for drinking, had permanent ice clinging to its southern shore. About 04:00 hours, a thin film of ice had formed on the surface of the entire lake, even though the sun had not yet set. It was, in fact, beginning to rise in the far northeast. I simply cannot aptly describe my awe, watching the sun cross the northern horizon, its nearly flat trajectory beginning as a slow climb back into the higher reaches of the southern sky. Only those who have seen this sight can fully appreciate it.

NASA astronauts would feel at home in the area around Ecstal Lake, as it resembles a moonscape. Shale covers the ground, and someone attempting to creep up on an unsuspecting person would fail because footsteps can be heard as far as a half-mile away when the wind is still.

The Americans were well equipped for their stay on the shore of the lake. A chemical toilet was housed inside a small tent, and they had built a shower using two forty-five imperial gallon drums with their tops removed. One drum contained cold lake water, while the other had a pair of blow pots mounted on a pedestal beneath, which heated the second drum of water. A valve system mixed hot water with cold, which allowed the bather his choice of shower temperature.

To the northeast, beyond the shoreline, rocky hills rose higher than the level of the lake; they seemed to block aircraft departure, but upon closer examination I could see that the rocks were distant enough to allow maneuvering room when departing the lake.

Just as I came to the water's edge, perhaps 200 feet above its surface, I was surprised and somewhat awestruck by the sheer drop of at least 1,200 feet straight into a fjord which ran close to 85 degrees to my departure path.

Back at Arctic Bay, several days following my first arrival, my fuel tanks were topped off and I filled the sixteen-plus gallon engine oil tank. I am sorry, in retrospect, that my stay at Arctic Bay was so short, because I would have loved to hear more of the ghost that haunts the Hudson's Bay Company buildings. I had no inkling of the story until several years later, when I heard about it on CBC radio. CBC hired an English parapsychologist to look into a report of apparitions that reportedly frequent the Arctic Bay HBO outpost. The investigator confirmed this was a well-documented haunting.

I was pleasantly surprised to discover CF-GTP had used only five gallons of oil on this trip so far. I was to take the two helicopter pilots and radio operator to Churchill. For the pilots, this would complete the first long leg of their return to Texas. The Canadian radio operator was a radio amateur as well, whose name I can't recall. Personal gear (including two sixteen-foot narwhal tusks) was loaded into the cabin of GTP. The amount of baggage left little room for comfortable passenger accommodation. One of the helicopter pilots occupied the right-hand seat on the flight deck with me; he made a gesture toward my instrument panel and asked where my navigation radio was. I pointed to my radio range receiver, which, of course, was absolutely useless here, since the nearest radio range facility was at Churchill. CF-GTP had an extra phone jack mounted on the right side of the instrument panel; I invited him to plug in and hear my conversations with the outside world. The other passengers were seated wherever they could find room in the cabin.

With almost no wind to ruffle its surface, Arctic Bay was almost glassy; our takeoff run was straight away from the settlement toward the opening at the south end of the harbor. We were loaded very close to maximum weight, probably on the high side, and when we reached the mouth of the harbor we had gained no more than 800 feet. I made a gentle turn 90 degrees to our right; I was again impressed by the sheer drop of those cliffs into the fjord.

The fjord carried us into Admiralty Inlet. I turned again, to the left this time, and set up a southerly heading, but without a specific number of degrees on my directional gyro, which would direct us to Repulse Bay. An overcast sky loomed above us with enough brightness to indicate a thin cloud layer. Bill Jensen had told me just after takeoff that Repulse Bay was experiencing broken cloud at 4,000 feet; with this in mind, I began to climb above the overcast in order to take a shot of the sun. At 5,000 feet we were above the cloud layer, in clear sky with an unobstructed view of the sun.

Magnetic north was displaced about 73 degrees west at our location, with so much vertical component as to render the magnetic compass totally useless, unless held stationary for at least two minutes to allow it to settle on the magnetic pole. I carried an extra watch, which I would set at local time (15 degrees per hour from Greenwich, England). The watch is held flat and a straight pin is stood on the crystal rim so as to cast its shadow across the watch face; the relative position of hands and shadow gives one the direction of true south.

This poor-man's sextant made it possible to set the directional gyro with precision. It allowed me to set up a heading that would take us west of Igloolik, directly to Repulse Bay. The use of this method of finding true directions may seem, in this day of Global Positioning System (GPS) navigation, somewhat crude, like using stone knives for brain surgery. The point is, however, after using the method a number of times I had come to rely on it; I staked my life on it.

Our heading and the elapsed time suggested we were approaching Repulse Bay; we were still at 5,000 feet and had a solid undercast deck beneath us. "Repulse Bay, GTP, we should be quite near you at this time, would you step outside and see if you can hear anything?" I radioed. There was about thirty seconds of silence. "GTP, Repulse Bay ... I hear an aircraft engine just west and to the north of here. I'm sure its you." He went on, "Our ceiling is 2,000, but I can see the hills clearly under the clouds."

"Repulse Bay, GTP. Ahh ... I'm beginning descent from five zero at this time ... GTP."

We broke out under the clouds about a mile west of the Hudson's Bay post at Repulse Bay. My Texas helicopter friend looked at me and just shook his head. Bill had had several Inuit men pay out the ropes during the night to accommodate GTP in the ebb and flow tide. He had no buoy close in for aircraft; only for the supply ship, several hundred yards from shore.

Our flight from Repulse Bay to Chesterfield Inlet was as normal as any. I was able to maintain visual flight rules and had no difficulty in finding our way. Although CF-GTP carried fifty gallons of fuel in both wing tanks, with thirty-seven gallons in the front belly tank and another sixty-four gallons in the rear belly tank, for a total of 201 gallons and seven hours flying, we had no choice but to land at Chesterfield Inlet for fuel. An east wind had sprung up and was increasing in velocity every minute. When we arrived at Chesterfield Inlet, the wind was blowing a gale across Hudson Bay. Fortunately, a natural rock breakwater extends south about half a mile on the east side of the bay. Those huge waves outside the breakwater were diverted; they rolled in 90 degrees to the wind.

Because of our heavy load, we weren't able to use the small lake at Chesterfield Inlet; we had no choice but to land parallel with those six-foot swells as they rolled into the bay. We skimmed the crest of one wave, went into the valley between two waves, then stopped on the crest of the next. An anchored raft holding several drums of 80/87 gasoline was the only thing I was able to approach. The drums were lashed to the raft deck, and I approached carefully, so as not to damage my airplane. I secured my floats to the raft and began the difficult task of transferring fuel, while the aircraft and raft pitched and rolled wildly in the heavy seas. I would not have landed at Chesterfield had I been flying any other bush plane; I would have flown inland and waited out the storm on some smaller lake. Such is the ruggedness of the Mark VI Norseman.

With the help of my passengers, we took on fuel from two forty-five-gallon drums, having pumped it into the two belly tanks for the rest of our flight to Churchill. My right foot was on the port float, the other on the buoy, when a particularly high wave arrived. That raft went down one side and GTP down the other. I heard a very loud report, like a shotgun blast, and I was suddenly aware that the ropes had parted. I had barely time to reach for the left strut as GTP was swept away. Our Pratt & Whitney R-1340 started at once, and I allowed it to warm up as GTP drifted tail first. She sailed toward the western shore. Aided by the gale of forty-plus miles per hour, our takeoff was a mirror image of our landing. We became airborne in no more than three lengths of the airplane and climbed with about a groundspeed of twenty-five miles per hour.

"Churchill Radio, GTP," I called, "Off Chesterfield at 16:20 Greenwich. Estimating Churchill at 22:08." Tex, as I now called my Texan passenger, studied his wristwatch with a look that suggested he didn't really trust my numbers, given the radio navigation equipment we had onboard. Five minutes following our departure, we encountered a layer of fog that became more dense with each passing minute. I called

Chesterfield Inlet. "Chesterfield radio, we're in fog here and wish to return Chesterfield." I had no idea where we could keep GTP safely, but this was not looking good.

"Negative, GTP ... fog rolled in here just after you lifted off. We are now at zero."

"Ahh ... okay," I said. "We'll continue to Churchill, or maybe we can get down at Eskimo Point. Thank you, sir, good morning ... GTP."

Hudson Bay weather is unpredictable at best, and because of it, all Visual Flight Rules (VFR) operations are in aircraft that are Instrument Flight Rules (IFR) equipped. "Churchill radio," I said, "GTP unable Chesterfield because of fog; we're climbing to flight level five zero on instruments, and we still estimate Churchill at 22:08 Greenwich Mean Time."

With our heavy load, GTP would climb no higher than 5,000 feet, and we were in solid cloud. In about two hours, Tex found a single hole in the undercast; he shouted, "I just saw a ship!"

"Churchill Radio, GTP," I said, "By Eskimo Point, at one three, flight level five zero on instruments, estimating Churchill 22:08 Zebra." (We often used the word *Zebra*. Today, *Zulu* is the word used. They have the same meaning.) Tex looked puzzled until I told him I knew the resupply ship was at anchor at Eskimo Point. It was the only ship anchored anywhere near our flight path.

The northwest leg of Churchill radio range crossed the western coast of Hudson's Bay about ninety miles south of Eskimo Point, and we were in the twilight zone on its right side, where I was just able to distinguish the Morse "A" from the "N" signal. We turned left, heading directly for Churchill radio range and staying in the twilight zone.

We were about midway between the west coast and Churchill when the small tic of the oil pressure gauge needle attracted my eye. It was only a small one, but it worried me. Five thousand feet below, I knew, a heavy sea was running. I doubted even this Mark VI Norseman could survive a landing in that heavy sea, yet there it was again. A quick drop in oil pressure, no more than a tic, but we had about thirty miles to go before reaching our goal. I hoped my oil pressure would hold.

"Churchill Radio, GTP," I called again, "What is your current weather?"

After a long moment I heard, "GTP, Churchill ... ceiling 1,000 overcast, visibility one mile to the north, three miles to the south, wind zero five six degrees at fifteen knots gusting to twenty. Go ahead your position."

"Churchill, GTP ... we're fifteen miles northwest descending from five zero on instruments, still estimating 22:08 Zebra."

I continued inbound, hoping my passenger had not spotted the oil pressure needle. I shut off the automatic volume control on the range receiver and reduced sensitivity to where the signal was barely perceptible. The signal grew as we approached the station, until it suddenly became very loud, then dropped to near zero, only to build again to a high level a moment later. I knew we had just passed over the "cone of silence." Directly above the radio range antennas, there was virtual total cancellation of all signals because of the phase relationships of various signal components. This signal cancellation is called "the cone of silence." (Finding the cone of silence pinpoints a pilot's exact position, directly above the station.)

I made a steep turn to the right and quickly found the twilight zone on the southwest leg outbound. We continued this heading five minutes, then I did a forty-five–degree left turn, following the new heading for two minutes; a rate one turn to the left for one minute took us back on a heading to the right side of the beam inbound, leaving about four minutes to "bracket the beam." This would take us directly to the radio range station, located on the east shore of Landing Lake.

Radio range stations consisted of a single nondirectional antenna that received continuous wave radio frequency energy from a transmitter, which could be voice modulated for weather information or simply transmitted as an unmodulated carrier. The antenna, which was usually a single vertical tower, radiated this signal in every direction with equal strength. Four additional vertical antennas received a second signal from a second transmitter. This second signal was radiated 1,020 cycles (Hertz) above the carrier from the omnidirectional antenna. The new signal was alternately switched between vertical antenna pairs in such a way as to produce alternate "figure eight" radiation patterns. The energy was switched between antenna pairs in Morse code sequence, as to produce the letter "A" (Dit Dah) on one pair of antennas and the letter "N" (Dah Dit) on the other pair.

By controlling the phase relationships of energy transmitted to each vertical antenna pair, the installers produced "legs" pointing in any desired direction. This involved varying lengths of transmission lines as phasing networks.

In the aircraft, the pilot could select which signal he wanted. A three-position switch marked "Range–Voice–Both" was used. The very narrow filter would either notch out that 1,020 Hz Range signal, while allowing virtually all of the voice frequencies to reach the pilot's ears, or the filter could be inserted in series with the output, allowing only the range signal to pass and preventing voice from being heard. In the Both position, range and voice could be heard simultaneously.

Because of phase difference between the received signals in the aircraft, there was a narrow (one to three degrees) "on course" signal, where both the Morse code "A" and "N" signals were of equal strength. On that narrow beam, the pilot could hear both "A" and "N," and they combined to sound like a steady tone, but if he moved to his right (either inbound or outbound), he'd hear one of those Morse code signals, against a steady tone of lesser strength. This area (which could be as many as nine degrees wide) was known as "the twilight zone." A twilight zone existed on both sides of the "on course" signal.

A pilot flying within a twilight zone was always to the right of the on--course signal, just as automobiles travel to the right of white lines on highways.

As we descended through 1,000 feet still in solid cloud, I asked Tex to be my lookout. I called, "Churchill, GTP, will you give me your ceiling again please?"

"GTP, Churchill, roger ... we're showing 1,000 overcast."

"Go ahead your latest altimeter setting," I requested.

"Churchill altimeter setting 29.38 go ahead," the operator told me.

I set my altimeter to match. "Check 29.38 ... thank you, sir, GTP." I told Tex I would go no lower than 300 feet, although I was worried about the flicking oil pressure needle. We discovered during subsequent checks that fourteen gallons of oil had been used on our return trip; that "tic" was caused by remaining oil sloshing around and periodically uncovering the pipe that delivered oil to the pump.

At 600 feet, I asked Tex if he could see anything. He didn't. At 400 feet, Tex could still see nothing. We had just gone through 300 feet and my hand was on the throttle when Tex sang out, "Contact, I see a lake." It might have been one of the most beautiful sights in all the world. Directly ahead was Landing Lake. We were lined up perfectly, on final. As the keels of our floats skimmed the water, Churchill came on to say. "GTP, Churchill ... check you down at zero eight."

Tex studied his wristwatch intently and shook his head unbelievingly, as we taxied to the dock.

This ended my first excursion into the Arctic, but I would find a way to return to that beautiful and mystic land.

14 Fired

Upon my return to Flin Flon, I was pleased to find that George Dram had successfully looked after our fish haul contracts using the Cessna 180. We made less money than normal because George was charging the fishermen seventy-five cents per mile for 1,800 pounds of fish, as per our agreement. Since our Cessna charter rate was forty-five cents a mile and we were making three trips for each 1,800 pounds, we should have been bringing in one dollar and thirty-five cents for each 1,800 pounds of fish hauled. Of course I would not cheat my customers by charging more than the amount we had agreed upon.

Almost immediately upon my return to Flin Flon, Wagner insisted that I lend Norseman CF-GTP to our Lynn Lake base. I was given a different Cessna 180 CF-HPW. George Dram and I continued with the fish contracts using these two Cessna 180s. I protested the loss of my Norseman, of course, but to no avail. Norseman CF-GTP was away from August 23 until August 29. During this time, George and I kept busy with our fish contracts.

One day, Stan Wagner called me on the telephone, complaining that we were not producing sufficient revenue from our fish hauls. My answer was that since he had forced us into using the Cessna 180s in completing our fish contracts, he was the fellow who had caused this reduction of revenue from our Cessna 180 operations. Wagner's solution was that we carry 900 pounds per load in the Cessna 180s.

"Can you carry 900 pounds to a load in those Cessna 180s?" he asked.

"Stan," I said, "You know as well as I do that carrying a 900-pound load in a Cessna 180 is illegal." I paused for a second and then went on. "Cessna 180s can handle 900 pounds with no problem; both George and I can fly a 180 with that kind of load, but it is illegal."

At this point, Wagner said, "Okay, do it."

"Where will we stand if a Department of Transport inspector catches us with that big load on board a Cessna 180?" I queried him.

"Oh, listen fella," he said, "I will stand behind you if that happens."

"Okay, Stan," I replied. "You put it in writing—authorize me to haul 900 pounds in these Cessna 180, so I have a copy to hand that DOT

Inspector when I get caught. Then George and I will be glad to haul that kind of load."

Stan Wagner was obviously angry at my suggestion and said that he couldn't give me anything in writing, but I'd be safe under his assurances. I turned him down and suggested the best way to bring back proper revenue was to give back CF-GTP to me so I could carry on with my contracts, using the Norseman as I had originally done. I did get GTP back, but it was only a matter of time until I was ordered to go to Lynn Lake to help them with their fish contracts.

All this time, my contact at Lac Du Bonnet was complaining about no flying, at least not nearly enough to keep him happy.

A few days after I had finished helping out at Lynn Lake, I was sent to Wabowden to help clear their backlog, and while I was there it came to my attention that my engine time had almost run out. I had only four hours left before it became illegal to fly CF-GTP. It turned out that Lac Du Bonnet had a weekly schedule from the Lac Du Bonnet base to several places on the east side of Lake Winnipeg, ending at Cross Lake and then returning along the same route to Lac Du Bonnet. Lac Du Bonnet told me that they had a charter to Cross Lake the following day; we would exchange airplanes and they would take GTP back to Lac Du Bonnet and do the change, giving GTP back to me within twenty-four hours.

I called Stan to suggest that since I was already at Wabowden I could fly to Cross Lake and trade airplanes for a day. I said I could get a revenue-bearing load into Cross Lake nearly any time.

I had already checked that a load was available for Cross lake. Stan Wagner would not hear of it. "No," he said, "There will be non-revenue flying involved with it, and these machines must pay their way." I argued, in vain, that at most there would be fifty miles of non-revenue flying involved.

Stan Wagner insisted that I fly GTP back to Flin Flon, and a new engine would be brought there by truck. "That's a really excellent way to generate non-revenue flying, Stan," I told him, "It's 250 miles to Flin Flon, and the entire flight will be non-revenue. Meeting at Cross Lake would be no more than fifty miles." There was no point in arguing. Stan Wagner had already made up his mind.

I flew back to Flin Flon, and a truck was dispatched from Lac Du Bonnet with the replacement engine. Just south of Dauphin, a tire blew out and the truck ended up in the ditch. A second truck was dispatched, and during the several days it took for the new engine to arrive, I had a crew of four men staying in the local hotel, all being billed against my base. CF-GTP and I accomplished absolutely nothing during this period. Cessna 180 CF-HPW was taken away when Norseman CF-GTP was airworthy again.

About this time I was engaged by some of the International Nickel Co. management people. I was to fly them around to various points. One such incident involved an International Nickel Company vice president. He and his entourage chartered Norseman CF-GTP for two or three days, visiting various sections of the Inco holdings, the area generally in and around Moak and Cook lakes. At the time, I was a smoker. My weed of choice was Player's fine cut cigarette tobacco, and I would always use Vogue cigarette papers. I prided myself on producing a well-packed smoke.

As we flew out of Thicket Portage toward Moak Lake on this day, I began rolling a new smoke. The vice president nudged my arm, asking if he might try his hand at rolling one. I handed him my tobacco pouch and rolling papers. The man began rolling a fairly well-shaped cigarette. When I commented on the great job he had done, he pointed out that he had not always been a vice president of a large corporation. In his youth he had been a smoker with little surplus cash; he'd always rolled his own cigarettes. He contentedly puffed away at his masterpiece and said it was perhaps the best smoke he had enjoyed since his youth. "I wish," he told me, "that I could still smoke the same type of cigarette today, but it just wouldn't look right."

Following their inspection of the Moak Lake facilities, we flew on to Cook Lake (now called Thompson) for another round of inspections of drill sites. The next day, the vice president and his men were summoned for a meeting in Winnipeg. I was told that tomorrow's flight would take us south. Although I was employed by TransAir Ltd., I was to be treated as a corporate pilot in that I provided aerial transportation for executives of International Nickel Company. This included an invitation to sit at the head table during the dinner. When the business part of the meeting began, I found a reason to excuse myself—I imagine, had I not done so, I might have been asked to find something to occupy myself while the meeting was in progress.

When coffee was later served, I got out my pouch and papers and began to roll a smoke. The vice president again asked if he might try his hand at rolling a second cigarette. Then, another vice president asked if he might try his hand at rolling. In a very short time, everyone at the head table, including the president of International Nickel Company, was enjoying my home-rolled brand of nicotine delivery. It may be they were trying to put me at ease, but I believe these men thoroughly enjoyed the experience.

From September 1 to September 5, Norseman CF-GTP and I were content with our lot. We were flying an average of five hours each day and generating the proper amount of revenue. At this point, I lost my Norseman once again. I was getting very frustrated. A de Havilland Beaver CF-FHQ was assigned to our Flin Flon Base, and while it was a pleasure to fly this high-performance airplane, I still needed a Norseman to look after my fishing customers. I had Beaver CF-FHQ from September 5 until September 19. Then I ferried her to Lac Du Bonnet, with a stop at Berrens River on the east shore of Lake Winnipeg to pick up a load of wild rice. When that load of rice was aboard my de Havilland Beaver, I realized that I would not be able to legally reach Lac Du Bonnet before official night. Next morning, due to high water content in the wild rice, the headliner in FHQ was dripping with condensation. Upon my arrival at Lac Du Bonnet, I boarded a bus for Winnipeg. I was surprised when I was met by Sid Smith, executive assistant to the president, R.D. Turner.

Sid asked where I would be staying until the morning flight to Lynn Lake, where GTP had been on loan. I told him that I had planned on staying at the St. Regis but that they had informed me, not five minutes earlier, that they had no room for me. As executive assistant to the president, I'm sure Sid Smith had much more clout than a lowly line pilot had. He offered to call the St. Regis on my behalf.

"Sid Smith here," he said. "A very important staff member is in the city and is badly in need of a room ... thank you." He hung the telephone receiver on its cradle and turned to me, saying, "There you are, you will be staying at the St. Regis."

We got into Smith's Oldsmobile and headed for the Airport Hotel. Upon our arrival, we went directly to the Starlight Room, to a table occupied by the president, R.D. Turner.

I was beginning to feel like a VIP with all this attention from such a high echelon within the company. As we were progressing toward our dessert, Sid Smith said, "Al, R.D. and I have been discussing you and your future with this company." Turner very seldom spoke, preferring to let Sid do the talking. He smoked his pipe and would nod periodically. "We have seen enough now," Sid went on, "to tell you that you have a great future with TransAir." R.D. Turner took another draw on his pipe and nodded his agreement. "You can go ahead and build an empire at Flin Flon."

Shortly after, Smith and Turner drove me back to the St. Regis, and I went up to my room. In the morning I took flight 103 directly to Lynn Lake to retrieve Norseman CF-GTP. Norseman CF-GTP had been on loan to Lynn Lake. Now, we were reunited once again. Within a few days of my return to Flin Flon, Wagner called me, saying he wanted me to come to his office in Winnipeg as soon as I could arrange it. I told him I'd been asked by Jock Cameron at Lynn Lake to help him with his backlog of fish but that I'd get there as soon as that job was complete. He said that would be okay so long as I came right after the Lynn Lake work was finished.

On October 8, 1958, our number one son Mark was born, and when he arrived I was in CF-GTP, on another fish-haul from Reindeer Lake to Lynn Lake. Louise was still in the hospital with Mark when I was able to requisition a seat on board our DC-3 Flight 104 from Flin Flon to Winnipeg. I felt badly that I couldn't even pick up Louise and my new son from the hospital. But our friend Art McKeil and his wife, Eleanor, went to the hospital in my stead and took Louise and Mark home.

Once I was in Stan Wagner's office, he told me, "Well, fella, I've got some bad news for you." He said, "Seems the RCMP put a report into the Department of Transport on you. The DOT has told me that I have to ground you for a while." These were obvious lies, because the Department of Transport doesn't operate that way; they write a letter to a grounded pilot stating the infractions that he has incurred. In addition, the RCMP always approaches someone who has broken the law, or anyone they believe has broken the law, with a summons for that person to appear in court.

"I see," I said. "Well, what will I be doing in the meantime?"

"Well, fella," Stan said, "we don't really have a place for you right now."

I looked him in the eye. "What you're saying is that I'm fired, right?" He moved in his chair a bit and responded.

"Well fella, 'fired' is a pretty strong word. You're suspended." I said, "How long is this suspension to be?"

"Well fella," Wagner told me, "it's kind of indefinite, but you won't be real soon coming back."

So much for the assurances of Sid Smith and R.D. Turner.

I knew part of the reason for my firing. When Wagner had taken my airplanes away and when he had so blatantly overridden my suggestion of an airplane exchange at Cross Lake, I had expressed my feelings quite vociferously in the presence of our office girl in Flin Flon. She was Archie Talbot's daughter Myrna Talbot, although she was so different from Archie, it was hard to accept that they were from the same family. She was Gordon King's fiancée, and she resented the fact that I had tried to fire Gordon following an incident in which he had lied about pumping my aircraft floats.

Myrna had called Stan Wagner, and Wagner reinstated Gordon, over my strenuous objections. I uttered a few more choice phrases regarding Stan's intelligence. I feel certain that my remarks were promptly relayed verbatim to him. Under the circumstances, it might be reasonable to assume that Myrna had told Wagner of my heartfelt opinion of him. I am convinced, however, that it was, in part, a holdover from our old problems when I was with Ontario Department of Lands and Forests. There were some people within the company who called Stan Wagner "the elephant" because he would never forget a wrong or a perceived wrong.

Just before I left Flin Flon, I appointed my pilot George Dram as temporary base manager. George had the authority to requisition space on board the DC-3 service to Winnipeg. When George was told of my suspension, he secured a seat on a DC-3 and flew to Winnipeg and then strode into Stan Wagner's office. "If A1 Williams isn't reinstated," he told Stan, "I'm going to start my own air service in the area, and I'll run you guys out of business." (I suspect that the word a stronger word was substituted for *guys*.)

Wagner just laughed at George's proposal and refused to consider my reinstatement. George handed in his resignation on the spot and went directly to one of the local aircraft dealers. He purchased a Stinson 108-3 station wagon and began hauling fish for his fisher friends out of Wabowden.

It was time for me to leave the employ of TransAir Ltd.; I left a bit sadder and a bit wiser. It had been interesting, but I had been in a constant battle while I was in Flin Flon. I had endured too many aircraft changes during those critical times, with too many face-offs with Ulysses S. Wagner.

TransAir soon lost business to its main competitor, Parsons Airways Northern Ltd. George Dram began hauling fish into Wabowden and, at the suggestion of Stan Elliott, paid one dollar to become a partner in the various fishing companies he flew for, thus avoiding the fines he'd been paying for operating without a charter. Business prospered for George. He founded Cross Lake Airways with his new charter. George then made good on his promise to Stan, as TransAir bases in Flin Flon and Wabowden failed. George shook my hand when I left TransAir and told me that, should I ever need a flying position, he would have an airplane for me to fly. We have always remained friends.

A short time after my termination from TransAir, I was invited to go along on a fish haul with Lorne Goulet. Lorne had replaced me at Flin Flon and we had flown together several months earlier. Lorne flew de Havilland Beaver CF-JEI out to Suggi Lake, which is southwest of Amisk Lake. When we had taken on the load of fish Lorne turned over the return flight to me; it was the last TransAir aircraft I ever flew. I remain thankful for Lorne Goulet's gesture of friendship and trust that day.

As TransAir base manager I had learned how a bush flying operation should be run. Observing Stan Wagner's poor management methods was an education that no business administration school could have offered me. I'm sure he hadn't intended to, but Wagner had done me a great favor when he dismissed me.

Bush and Arctic Pilot

15 Custer Channel Wing

Two days following my leaving TransAir, I ran into Joe Richaud, who had been assigned to me as base engineer for a couple of weeks in Flin Flon. When Joe heard that I was no longer with TransAir, he immediately began talking of a new type of aircraft that had been built by Willard Custer in Flagerstown, Maryland.

Willard Custer was the great grandson of General George A. Custer of Little Big Horn fame. When Willard was young, he had been employed as block man for one of the farm tractor companies and posted in Kansas. One day, while servicing a disabled tractor in a Kansas field, he had to run for a nearby barn when a tornado suddenly bore down on the farm. Custer watched the twister from the barn door, thinking he was safe. As it advanced, the tornado veered suddenly toward the barn. Willard could only watch in horror. The tornado tore the hip roof from the barn and then carried it nearly 1,000 feet into the air; at this point the roof flipped over onto its back and glided, much like a paper airplane. Following a fairly flat trajectory into a neighboring field, it was destroyed on touchdown. The experience gave him food for thought.

Years later, Custer became an aircraft design engineer for Fairchild Aircraft Company in Hagerstown. Custer had lost two or three test pilot friends during takeoff accidents. He wondered why any aircraft must reach dangerous speeds while running down a runway for takeoff. His mind drifted back to his days as a block man and the day the tornado ripped the roof from the barn.

While the barn roof really didn't have an aerodynamic curve to its upper surface, it had created some lift, much like an airplane folded from paper does. Custer began thinking that an airfoil should not necessarily have need to move through the air. What if the air was made to move over an airfoil shape. Would there be not be lift?

Willard Custer reasoned that it was the shape of the barn roof which must have given it a glide ability. He set out to prove or disprove his theory.

He left Fairchild Aircraft in Hagerstown to devote most of his time to developing an airplane around "channel wings." He bought out the defunct Bauman Aircraft Company, which had built the Bauman Brigadier. To this twin pusher engine aircraft he fitted his channel wings and designated the aircraft as the CCW-5 (Custer Channel Wing, model five).

Meanwhile, Joe Richaud contacted Paul Sigardson in Winnipeg, who, together with associates, founded Channel Air Ltd., which was to be a sales company. Joe would be the engineer on the project while I was to be demonstration pilot. We were told that Custer was to manufacture the new aircraft in the Montreal Noorduyn factory, where Noorduyn had built his Norseman aircraft.

Before we began selling the aircraft, we needed to obtain a contract with Custer Channel Wings Ltd., which was now known as Custer Channel Wings of Canada Ltd. I flew to Montreal and contacted Bill Spence—the guiding hand for the company. Bill Spence approved of our efforts in promoting the new aircraft. He told me that a contract would be drawn up and mailed to us in Winnipeg. Richaud and I drove to Montreal prepared to learn all there was to be learned about this revolutionary new aircraft. When we arrived in Montreal, we discovered Channel Air Ltd. hadn't signed the contract.

Financing had proved elusive, and CCW-5 production was never begun.

The technical aspects of the aircraft were above reproach, as I was to discover first-hand. Spence and I flew into the Akron/Canton Airport in Ohio in his Beech 18 during the early months of 1959, while Joe drove there with my 1955 Ford and a newly designed engine mount for the aircraft.

I saw the CCW-5 in action for the first time in Ohio. With his brakes set, the test pilot opened up to takeoff power. After dragging its wheels for about fifty feet, the CCW-5 became airborne—with its wheels still locked. It lifted off at an estimated forward speed of only eleven miles per hour and climbed out at an angle of about 45 degrees.

I was saddened when the project died, but I needed to make a living for my family. I returned home to find a flying job.

16 Looking for Work

Whenever I applied for a flying position, I was turned down. I eventually learned I had been blackballed by my old nemesis, TransAir's general manager, Stan Wagner. I discovered the truth when I applied for a job at Taylor Airways, owned by Jim Shore.

Jim Shore had made his mark when he established Office Overloads, a company that offered a pool of experienced office workers. Shore also owned Teal Airways, which operated out of Winnipeg. He telephoned me from Winnipeg and asked that I meet him at the Regina Airport the next day. (We were in Moose Jaw at the time.) Jim arrived in Regina at 14:00 hours to interview me for a pilot position. Over coffee, he told me that I was hired; he went on to tell me that when he had checked my references, he was advised not to hire me.

Stan Wagner had said so many negative things about me that Shore had decided to check me out for himself. "I pride myself on judging character," he said. "This man painted you so black that I just had to meet you and see if anyone could be as bad as you were made out to be." I told him earnestly that he wouldn't regret his decision. Shore looked at me carefully before he replied. "If I'm disappointed, it will be with my own judgment. I'm satisfied there will be no disappointment. By the way, when you apply for a position in future, I suggest you do it in person. That will almost certainly secure you the job without regard to what some outsider has to say." I was flattered by his comment.

Shortly after our conversation, I went to work for Taylor Airways to fly Piper Super Cub 135 CF-JFO. I flew Cessna 180 CF-JEN to Wabowden; I have heard the 135 Super Cub was a much better performer than its 150-horsepower twin, Super Cub 150, but since I only flew the 135, I can't comment on their relative merits. The Super Cub 135 certainly was a fine performer. Her weight and power made her agile and a delight to fly. She was virtually impossible to overload, due to her relatively limited interior volume.

On one flight, I had a passenger who weighed about 275 pounds; he had a pack and bedroll that added probably a hundred pounds more.

He carried a Winchester .44-.40 rifle, 100 rounds of ammunition and two cases of twenty-four bottles of Molson's Canadian. His plan was to stay at his cabin, about forty miles north of Wabowden, during spring breakup. He was well known and liked by a good many people, some of whom were at the lake to see him off.

Probably less than a quarter-inch of melt water lay on top of the ice when we had loaded JFO. Winter flying season was coming to a close, and my Piper Super Cub 135 was the last aircraft operating in the area; all heavier aircraft had returned south. After warm-up, during which JFO never moved an inch, I popped a few degrees of flap and opened the throttle for takeoff. When I returned, several persons who had witnessed the takeoff were still at the lake. They said they had measured my run from start to becoming airborne and that I had used only 103 feet for the entire takeoff. I know that the thin coating of water would certainly have helped, but it was still a surprisingly short distance, considering the heavy load the Super Cub had carried.

By nature, I have always been a person who stayed on friendly terms with my competitors. Stan Elliott, the flying base manager of TransAir's Wabowden operation, and I were on good terms. Elliott was a good man in anyone's book.

Just before Stan took his Beaver CF-JEI south for the change over to floats, he handed me his 8-mm movie camera and asked me to take it along on one of my Super Cub flights. Stan was especially interested in getting a movie shot from my Super Cub doing a full loop. I did a loop after climbing to 3,000 feet, with the camera tied to the cross struts in the windscreen of JFO, looking forward just above the engine cowling. I pushed over into a dive and built my speed to 135 miles per hour; then I pulled up into a vertical climb, went over the top, reducing power as I came down the back side of the loop. The camera was running during the entire process, from level flight to my dive and then all the way around the loop. I then returned the camera to Stan. It was well into our summer flying season before I had the opportunity of viewing my film. I took Norseman CF-DRD from Eldon Lake to Wabowden. I went to visit Stan and he showed me the movie.

That film is probably still being shown at home by Stan.

While performing mild acrobatics can be fun for a pilot, common sense must be applied to flying. Even good pilots like Stan Elliott and Al Nelson (Al and I had flown with Rusty Myers' Flying Service) can make errors in judgment. On this occasion, they both showed a lack of common sense. A commodity which even the best bush pilot won't live long without applying to his job.

Elliot and Nelson were both employed by TransAir Ltd., in the Cook Lake and Moak Lake areas. On this particular day, Elliott was flying de Havilland Beaver CF-ITW and Nelson was in Cessna 180 CF-IRQ.

Elliott was returning to Thicket Portage for another load, and Nelson was on his way into Cook Lake. Elliott sometimes would read comic books during his trips, looking for other traffic every few seconds as he went. On this trip, Stan looked up from a comic to see Nelson's Cessna 180 CF-IRQ directly in front of him, standing on its left wingtip. Elliott dropped his comic, which went sliding across the cockpit. He grabbed the control column and cranked in full left aileron as the two aircraft zoomed by each other at the combined closing speed of nearly 240 miles per hour, their floats no more than three feet from each other.

When the day's flying was completed, both pilots arrived back at Thicket Portage. Nelson approached Elliott and said, "Boy, you play one tough game of chicken. You hang on to the last second!" Elliot angrily explained that he had not been playing chicken or any other brainless game. He then went on to question the intelligence of whoever had granted a pilot certificate to Nelson. Stan and Al got into a shouting match and it seemed that fists might start to fly. In the end, however, Stan settled for making Nelson promise not to play chicken and to pay attention to the job. For his own part, Elliott read comic books only when his feet were firmly planted on the ground.

Breakup was fast approaching, and lake ice was beginning to candle (the ice had fragmented and fragments grouped together to form thousands of vertical candles). It was May 9, 1959, and flying would soon be over until after breakup. Sven Peterson called me on his Sunair S-5-DTR transceiver on that May morning. This two-way radio was used in most of our aircraft at that time; he had purchased one for his own use because he had only a twelve-volt electrical system at his camp and the Sunair was very easy to hook up to that system.

"Good morning, Sven," I said, "Can I help you today?"

"Well," he said in a heavy Swedish accent, "I think I got problems here. My ax slipped the other day. I cut my leg a bit, but now it's lookin' pretty bad."

I thought briefly about my own safety in flying during breakup. "Will you need a pick-up, Sven?"

"Yaw," he said, "I think I need a doctor to fix up my leg all right."

Recently, in Montreal, where I had gone to view Custer's CCW-5, I had met George Mah. He and his brother Cedric had flown over "the hump" in Burma during the war. George told me how he had operated from open water one whole winter, on ski-equipped aircraft. He told me he'd run into no problem. Skis would plane along the water at quite a low speed. Ice or snow on the shore was all one needed for a safe operation. Today, I was going to prove or disprove George Mah's story.

"Okay, Sven," I said into the microphone, "Tell me about your ice conditions on the river."

"Well," Sven replied, "we got wide open water on the river." His cabin was quite close to the river, and because the river widened at that spot, there was a flat shoreline between his cabin and the water's edge. I asked, "Do you have any ice or snow on the shore between your cabin and the water?"

"Yaw, the snow is about two inches tick and goes right down to the water."

I took off with JFO, wondering if I could pull off this rescue. A botched rescue operation and my own demise seemed distinct possibilities. But I had to try. Blood poisoning can be very serious. I was the only one who could attempt the rescue, as I had the only aircraft within at least 500 miles—even the single helicopter had gone south.

When I arrived over Sven's cabin, I surveyed the situation. The entire river was open water. 1 dragged the river, looking for a log or other landing hazards, but the water was open and clear.

I lined up for a landing parallel to the shore. I would need all the room available to try my first landing attempt. I touched down gently on the surface not far from the approach end and began slowing. At an estimated fifteen miles per hour I began getting nervous; there wasn't much room left. I opened up the throttle and left the surface. So far, so good, I thought.

I came around a second time, still parallel to the shore, but this time I slowed even further. At about five or six miles per hour, my skis began to porpoise, which gave me a fright. I added just a touch of power, which caused the skis to stop dancing. Once back in the air and knowing that George Mah had not lied, I lined up for Sven's front door. I reached his snow-covered beach and came to a stop no more than fifty feet from Sven's door. Our takeoff went smoothly. My skis seemed to be the perfect "feet" for this kind of water operation.

The following day, I was asked to pick up some men who had been doing line cutting for a proposed highway to Thompson. Unless I picked them up soon, I wouldn't be able to land and they would be forced to walk back home to Nelson House. My wife, Louise, and son Mark were visiting her sister in Medicine Hat then; she knew I would be coming out of the bush very soon, because warm weather was about to end winter flying for another year.

I made the first of two flights without incident; on the second trip I ran into snow flurries. As I turned toward Nelson House, the snow became a bit heavier. Soon visibility was reduced to less than a quarter of a mile. I arrived at Nelson House in blinding snow, and because the snow was very wet I put wing covers on JFO. I was welcomed by the ladies at the nursing station, who made up a room because it looked as though I might be a guest for longer than I had expected. I asked them if I could use their radio.

I was unable to raise any station, but I was told my message would probably get through via the Hudson's Bay Company radio network.

I had no luck with the HBC network, because signals were out. We were completely cut off from the rest of the world. Jim Shore held off as long as he could, but when two days had gone by with no word from me, he began thinking I had gone through the ice, or worse. He reluctantly telephoned Louise in Medicine Hat to inform her that I had disappeared.

"It's too soon to tell if Al has had any real problem, and I'll keep you posted; I thought I better let you know the way things are right now," Jim told her. Naturally, Louise began to worry. With a new son to look after, she didn't look forward to raising him with no father. The next day, the sky cleared and I was able to leave Nelson House after removing about ten inches of snow from my wings. I'd been kept comfortable, and the nursing station staff had fed me better than any stranded person could have expected.

Just twenty miles out of Wabowden, I finally was able to make radio contact, which shows how poor radio conditions had been for the past few days. RCAF Search and Rescue aircraft were preparing to leave Winnipeg for the Nelson House area when I made contact. The planes had to be called back from the departure end of the runway.

I departed Wabowden shortly after and arrived in Winnipeg for a few days off, then traveled on to Medicine Hat. I was asked to work in the hangar until breakup was complete, to make sure all the radio equipment was working properly.

I didn't much care for working in the hangar; I couldn't see out of the building unless I walked to the doors. I felt like a prisoner and longed to climb into an airplane, where I could view the landscape from a hundred miles away.

Break-up arrived soon enough, and I was introduced to Norseman CF-DRD. We would spend many happy hours together. DRD was fitted with wheels when we first met; I had never flown a Norseman on wheels before. I refueled before flying her out to a strip on the Red River, north of Winnipeg, where Konnie Johannesson and crew were to fit CF-DRD with floats for her summer operations.

1 taxied to the gasoline pumps, gently applied brakes—and was shocked. CF-DRD, like all Norseman Aircraft, was a bit nose-heavy on wheels. It was recommended that at least 100 pounds of sandbags be carried behind the cabin doors for ballast, to prevent her from standing on her nose when braking. Nobody had told me, therefore I didn't know about that, and so I released the brakes immediately and taxied around for one more approach to the gasoline pumps. This time I taxied much more slowly and used a very light touch on the brakes. Once I had her refueled, I flew Norseman CF-DRD out to the strip for change-over. I was now ready to go north for a summer of float flying.

A. R. (Al) Williams

17 Eldon Labe Operations

Norseman CF-DRD and I were to spend many happy hours in the air together flying out of Eldon Lake. This lake was the fish packing and bush flying center for the community of Lynn Lake.

Sherritt Gordon Mines ran their Company Air Service out of Lynn Lake, about three miles from Eldon Lake. We shared the general flying business with TransAir Ltd. Hank Parsons operated his secondary base there but concentrated more on pilot training than actual bush flying. Fred Chiupka operated his two Norsemans on his own fishing operations, so they were not really our competitors. They hauled their own fish to market. They did not, nor could they, compete for other fishermen's business.

Fred had hired Paul Ricky, who had been a TransAir pilot, and in turn Paul had hired recently retired RCAF pilot Chuck Saba. Paul had flown for TransAir a year earlier. One day at Lynn Lake, Paul had offered me the chance to fly the company's Bellanca Air Cruiser CL-BTW. This was a very large aircraft that utilized a Pratt & Whitney hornet engine rated at 850 horsepower. The airplane could handle about 3,200 pounds of fish to a load.

Paul Ricky told me that the big Bellanca had not flown for four weeks and that the floats had taken on a good deal of water in the step compartments. If I would pump those floats, Paul told me, he would give me a check ride in her. Because CF-BTW was the last of her kind still flying, I wanted to fly her. The day was hot, and after I had pumped most of the water from her port float, I was very sweaty. I elected to put off the balance of the pumping exercise until a cooler day. Unfortunately, the weather did not cool off while I was at the lake; I never had another opportunity to fly the Bellanca Air Cruiser. To this day, I regret not having seized that once-in-a-lifetime opportunity.

I arrived at Eldon Lake on June 15, 1959, and immediately began routine bush flying out of our Lynn Lake base. I ended that day at 22:10 hours, after putting in six hours in CF-DRD. That Norseman and I were fast becoming friends, but only after our engineering helper George had plugged the many openings in her fire wall. An almost unrestricted engine heat had poured through the numerous openings onto the flight deck. The following day saw me in the air another six hours. This was going to be a busy summer, and I was looking forward to every new day of flying. I flew to Arctic Lodge on Reindeer Lake, Granville Lake, Laurie River Power Dam, Cormorant Lake and Southend (at the south end of Reindeer Lake). It felt so good to be back in the air with a real airplane.

I had flown Piper Super Cub CF-JFO and Cessna 180 CF-JEN, and just before leaving the TransAir position, I'd flown de Havilland Beaver CF-FHQ for many days, but none of those aircraft reflected my personality as did the Mark VI Norseman. My current mount, CF-DRD, was perhaps the best of all Norseman which I had flown. I felt we had been destined to fly together.

On June 28, late in the afternoon, I was called upon to go into Brochet; I was to pick up a number of men who would work at Arctic Lodge—on Dewdney Island about fifty miles south of Brochet. Darkness was rapidly approaching as we departed the lake at Brochet at 23:14 hours. When our floats left the water, visibility was good and we had a routine flight to Dewdney Island, where I remained for the night. Next morning, I returned to Eldon Lake and immediately was confronted by an RCMP officer. He informed me that he had been at Brochet the evening before and was now charging me with taking off from an unlighted airdrome at night. I didn't agree with the charge and told the officer that I had made a legal takeoff.

Almost immediately afterward, I left for a fish pick-up flight to Murdo Clee and his brother's fish camp on southern Indian Lake. The settlement of South Indian Lake was at least two hours from Murdo's camp by boat, and the fish camp had no radio. Upon my arrival at their dock, only an hour and thirty-five minutes after being charged by the police, Murdo greeted me. "We hear you are in trouble with the law. My brother (he never called his brother by name) and I have talked it over. If they find you guilty, we want to pay your fine." I was surprised.

"How did you find out about that?" I asked.

"Moccasin telegraph," I was told. I tried to get more out of Murdo Clee, but I could get no further explanation. There was simply no way that this information could have been received from the South Indian Lake settlement, as they had no two-way radio at their fishing camp. To this day, I don't have any idea how they knew the RCMP had laid a charge against me. The only reason I can think of for their noble offer to pay my fine, outside of their being good neighbors, was that I treated them as the valued customers they were. They were not treated in that manner by everyone who dropped in. I have actually overheard a pilot say, "Come on, you goddamn bows and arrows, let's get this airplane loaded so I can get out of here." I found this attitude disgusting. The Cree were helping to provide an income for that particular pilot, yet he seemed to look down on them.

Murdo Clee and his brother were at least partly responsible for the total self-sufficiency of the southern Indian Lake community, now known as the O-Pipon-Na-Piwin First Nation. I held them in high regard, not only as customers but as friends.

Murdo did have his problems, however, with alcohol. One day when Murdo was in Lynn Lake to pick up a few supplies, he became sidetracked and showed up at our Eldon Lake dock ready to fly home. He was drunk, and as we settled down on course for his fishing camp, Murdo began poking his fingers at various switches and controls of CF-DRD. I couldn't reason with him, so I propped my right arm against the right side of the instrument panel and planted my right elbow firmly in his chest. We arrived at his camp with no problems. Poor Murdo complained for two or three days about chest pains suffered from bruises he'd received somewhere. I never told him that those bruises had come from my elbow in his chest.

At the urging of the other pilots, I sought legal counsel over my RCMP charge. My lawyer, Evans Premachuk, was not pleased when I told him I had made a statement but cheered up when he realized that I had not made an incriminating statement. Evans was articling for his law degree and was admitted to the bar a year later. I was to be his first court case.

Premachuk helped his mother in their general store, and he would periodically travel to Winnipeg on buying trips. I told him that I was not guilty of taking off from an unlighted airdrome at night, since the policeman had misunderstood the air regulations. The cop had charged me on his belief that night in Canada began one half hour after sunset. In fact, the regulations state, "Night in Canada shall be that period of time between evening civil twilight and morning civil twilight and shall begin not less than one half hour after sunset and shall end not less than one half hour before sunrise."

I convinced the lawyer that I had a case. He offered to check with the astronomy department of the University of Manitoba and the air force to see if they could confirm when night officially began in Brochet on the night of June 28. Latitude of a location plays an important role in when evening civil twilight ends, the lawyer agreed. Brochet, at the north end of Reindeer Lake, is near latitude 58 degrees north at 101 degrees 35 minutes west.

The log of Norseman DRD would have recorded the actual time of departure, while my pilot log only records the amount of my flying time and place of departure. It was very near 23:14 hours, in any case.

The university astronomy department and the Royal Canadian Air Force agreed within a few seconds of each other. Evening civil twilight, they told Evans, ended at Brochet on June 28 at 23:22 hours. They suggested that I enter a plea of not guilty. Unfortunately, the RCMP could not produce their star witness, an operator with the Department of Transport at Brochet, who had witnessed my departure together with the police constable. The magistrate came down with a ruling to postpone the actual trial until December 23, just two days before Christmas.

Evans asked who would pay for my transportation to Lynn Lake—I would be in Saskatchewan at the time of the trial. "Oh. the defendant will pay his own way," the magistrate said.

"What happens," Evans asked, "if my client refuses to return?"

The magistrate looked as severe as he could. "In that case," he told Evans, "your client will be arrested and returned under restraint."

"Well," Evans asked, "who will pay for his transportation in that case?"

The magistrate did his best to look even more stern. "Transportation would be covered by the Crown," he said.

"Well," Evans Premachuk said, "I shall advise my client not to answer any summons, and it will be the Crown's responsibility to have him transported here at that time."

This charge and the fear of losing my license hung over my head like the sword of Damocles during the entire summer of 1959. I must say that Jim Shore stood solidly behind me. He offered as much support as was within his capabilities. He told me that if a charge was laid against me, "we will back you up 100 percent." I have always been grateful to Jim Shore for that support. He is, in my books, the best. It was December 22, 1959, when I received a call from my lawyer.

"The justice department just this morning has thrown out your case, after checking with the astronomy department in Ottawa; they have discovered, as we did, that you are not guilty; with the result they have put a stop to this farce." Evans was a happy man, because this was his very first case and he had won it, hands down. A good start for a new defense lawyer.

It was about this time when I received a telegram from Pacific Western Airlines Ltd. PWA had an opening for me in their northern bush-flying operations. It was an attractive offer, but I felt honor-bound to Jim Shore for his decision to hire me when I was blackballed.

There are those who might say that I was foolish, because a number of other pilots who began with PWA at that time went on to become captains on DC-6s and then graduated to Boeing 737s. One former bush pilot even moved up to captain a Boeing 707. I completed my year with Taylor Airways and never regretted it. Poor Jim Shore was killed when his Grumman Widgeon crashed in the God's Lake area during the Pan American Games in Winnipeg in 1967.

Before the RCMP charge was dropped that year, there were many days when we began our flying day at 04:30 hours. Often my day would end about 23:00 hours, just the kind of flying I loved. Bush pilots in those days were the only truly free human beings on the planet.

One morning, I arrived at our Eldon Lake Base very early in the morning hours and prepared to depart for Jackpine Island. Ingvar Stolberg, who operated that fish camp, had a backlog of fish. It was imperative that we get his fish to market before they spoiled. I took off in semidarkness and didn't notice the fog that covered Eldon Lake until I broke out on top of a heavy layer of the white stuff. It would have been suicidal to try landing back on the lake from which I had just departed, and my fuel tanks were filled, with about seven hours of flight time.

The sensible thing to do was continue to Jackpine Island in hopes that this morning fog might burn off and I would find a safe haven with Ingvar Stolberg's crew. As I was coming up on the Paskwachi River, I could see a number of trees standing atop a hill off to my right. The hill was not visible, but I could see about the upper ten feet of the forty-foot-tall forest. This gave me hope that I would be able to land at Jackpine Island in about twenty minutes.

I called in to Jackpine Island. "Jackpine Island, DRD," I called. "By Paskwachi River at two zero ... estimating Jackpine at forty-two ... DRD."

"DRD, Jackpine." It was Ingvar's Norwegian accent that answered on the radio. "You can't land in here, we are socked in with heavy fog."

"Jackpine, DRD," I replied, "I've got fog all the way back to Lynn Lake, but I have plenty of fuel, so I'm coming in to take a look."

When I arrived over his camp, Ingvar Stolberg called. "DRD," he said, "We can hear you overhead, but we can't see any ting more than fifty feet from the dock."

"Okay, Ingvar," I said. "I can see the very tall tree beside your building, and I also see the tops of trees on the island straight out from your dock ... I can safely land between the trees, because there isn't any obstruction out in that stretch of open water ... DRD."

My landing was not a problem; the only trouble I had was in trying to find the dock. I turned 180 degrees and taxied back as far as I thought was reasonable to have passed the dock. Then I turned to my right and very slowly taxied toward the eastern shore of Jackpine Island. I could see the shoreline about seventy-five feet away, and I followed it north until the dock system loomed up out of the mist.

We loaded Norseman DRD, and I took off from that fogbound open water in front of the camp; I turned on course southeastward toward Eldon Lake, our Lynn Lake base, in hopes that the fog would have dissipated. When I arrived over Eldon Lake, I found that although there was an overcast sky above the fog layer, I could see straight down, and beside that, I could see the radio masts standing at least twenty feet above the fog.

I landed and taxied directly to the fish packing house to unload. As soon as that was done, I taxied out for takeoff again. I still carried plenty of fuel for at least two more trips. I didn't know it at the time, but

Eddie Dyck was upset; he raced over to the fish dock and arrived just as I was taxiing away. Ed had a reputation of flying through weather that would ground most other pilots. Today, however, because of the fog he was staying on the ground.

Looking straight up, it seemed to him that it would be impossible to fly because of overcast above the fog, which made it appear that the sky was solid overcast. He wanted me grounded safely for the rest of the day.

On the second trip, I taxied to the fish dock and unloaded. I learned later that Eddie took off with his 1956 Lord Customline for my position at the fish house. The road between our base and the fish dock was long and rough; it had been hacked out of the bush, following the path of least resistance. Eddie arrived at the fish dock just after I had pulled away the second time in a row.

On my next flight, during which visibility was improving a bit from air to ground—but not in the reverse direction—I pulled up to the fish dock and unloaded. Eddie heard me and came belting out of the Taylor Airways base. When he arrived at the fish dock, I was in the process of taxiing to our dock for more fuel. Eddie took off again for our dock and arrived there just as I was departing once more.

I finally arrived back at Eldon Lake, finished for the day. Eddie Dyck had given up; he had gone home for the evening. Louise and I shared a bathroom with Eddie and his wife Frieda, who lived in the front of our front/back duplex house. She was in the bathroom and heard Eddie complaining about how he had not been able to contact me; our office radio had a problem, which I had not gotten around to repairing—I did all of the radio maintenance at nearly every base where I worked.

Louise was furious and confronted Eddie about his remarks, telling him that I had been blackballed by Stan Wagner, that I had been given a second chance by Jim Shore, and that I was doing my best to justify his faith in me. If I was flying, she told him, I was doing it in a safe way, even if that didn't seem true from a ground's-eye point of view. Eddie backed down, and I never had any harsh words with him on the subject. I did explain how it was that I was able to continue moving Ingvar's fish that day. I made many trips that day, and mine was the only propeller to turn during the entire day.

I liked Eddie Dyck and I admired his flying ability; I believe he usually reciprocated my feelings. We certainly got along like two good friends throughout our association. Our close association almost ended in an accident which would have been the result of my own foolishness.

The prudent pilot never, under any circumstances, attempts flying in formation with another aircraft unless he has made previous arrangements with the pilot of that other aircraft.

Nobody had told me this fact, and I had not yet figured it out for myself. Eddie Dyck had taxied Norseman CF-DRC to a position not far north of the point of land at Eldon Lake, where our base was located, along with a scattering of other buildings, which included the TransAir Base and several fish-packing houses.

The wind was about fifteen miles per hour from the southeast, so Eddie's takeoff run would eventually carry him to within about 200 feet to the left of that point of land. I lined up behind him in Norseman CF-DRD and to his right at a distance of perhaps 100 feet. I opened my throttle almost in sync with his own throttle opening.

Both our Norseman aircraft were on the step when Eddie noticed a change in the wind direction—it had come around more nearly to the south; he changed his takeoff path accordingly. This was to bring him much closer to that point of land and it placed me in danger of running aground, so I was forced to move into a line astern position behind him. If Eddie had known I was there, he would have maintained his original takeoff path.

Being forced into line astern position, my Norseman was influenced by turbulent air behind his airplane. As we both became airborne, CF-DRD rolled to her right, forcing me to maintain full left aileron to prevent her from digging her starboard wingtip into the water. Once we had attained about fifty feet of altitude and were above trees and antenna masts, I was able to move to the right and climb without further turbulence from the wake of Eddie Dyck's Norseman. I never tried a formation flight with another aircraft, unless both pilots were in agreement with the flight before it was attempted.

On July 17, 1959, I was flying CF-DRD to South Indian Lake, carrying a load of cargo for the Hudson's Bay Company outpost. We were just coming up the west shore of Hughes Lake when a cylinder head blew. I made an entry in my log book: *H' Wasp blew 'Jug.'* Fortunately, we were very near Hughes Lake's west shore, with plenty of room to land.

I moved the fuel selector to off and stopped the flow of fuel; I also switched off the magneto switches, then reduced throttle and moved the propeller pitch lever settings to full coarse pitch so as to eliminate most of the propeller drag.

The landing placed us close to the eastern shore, and I was able to paddle Norseman DRD into a safe haven on a small point of land, which provided a sandy beach with plenty of timber nearby for our crew to use as materials. They would be needed to build an A-frame when the engine was being replaced. In response to my radio call, Eddie Dyck came out to Hughes Lake to pick me up with Norseman DRC, and in due time arrangements would be made to bring in a crew to make an engine change on CF-DRD. Since Eddie was paid a straight salary and I was paid much less salary but with mileage pay, Eddie Dyck gave me his Norseman DRC until DRD was back in service, a very generous offer for Eddie to make, it seemed to me. Before I could make any trips in DRC, however, I had to fly to Winnipeg. Engineer Henderson and I flew to Sioux Lookout with Cessna 182 CF-LBO.

This was my first encounter with the use of an omni-range navigation system, and I learned all I needed to know in just a few minutes. Our trip was for me to pick up Cessna 195 CF-DFD, which was needed at Eldon Lake in order to carry on our trade and with which our agent John Foster would build some flying time—he had just obtained his commercial pilot's certificate.

I flew out of Sioux Lookout on July 21, 1959; my first stop was at Red Lake. As district operator for the Department of Lands and Forests, I had worked Red Lake many times each day via radio, but I had never landed on Howey Bay—the fact is, I had never visited Red Lake during those years.

After refueling at Red Lake I flew to Lac Vert to pick up Louise and our son Mark for transport to Lynn Lake.

To date, I had only landed on Lac Vert; I had not tried to take off from the lake. (Its true name is Porterfield Lake, but everyone refers to it as Lac Vert Lake). Since I had very limited experience with the Cessna 195 on floats, I was unsure how well she would take off. I asked my brother

Garnet to act as guinea pig for a test flight, as Garnet weighed more than Louise and Mark combined.

Once we were in the air we flew without a quiver in the fairly quiet Cessna 195. Garnet, who obviously was thrilled with the flight, asked: "Do they really pay you for doing this?" he asked. We made a fuel stop at the Pas; then Louise and Mark and I flew on to our Eldon Lake base.

Following a few trips with CF-DFD, Eddie asked me to give Johnny Foster his check out in the 195. Poor Johnny, his first solo in DFD was to be his last in 195; after a good steady takeoff and climb, he turned downwind when suddenly his propeller went into full coarse pitch and he was unable to continue his circuit. He tried turning in from his downwind leg and landed without incident. It was a long taxi to shore, and I stood on the dock wondering what had happened to make him cut his circuit short.

Johnny was not injured, and we were not able to discover what had caused his propeller to go fully into coarse pitch. The last time I saw CF-DFD, she had been pulled partially up on the west shore of Eldon Lake with ice forming around the heels of her floats. I don't know if she ever flew again. She was probably repaired the next year, but I never heard about it.

At this point I began using CF-DRC—sister-ship to DRD—and on July 28, 1959, just a week after the blown jug on DRD, I was on a flight to Ingvar Stolberg's camp on Jackpine Island for a load of whitefish. Just as I was coming up on Paskwachi Lake, a place of about two miles width on the Paskwachi River, I smelled smoke. I had a fire onboard.

"Lynn Lake, DRC, I'm landing on Paskwachi Lake ... I have a fire onboard ... I'll call you when I'm down." I tried to sound as nonchalant as possible, but I must confess that I was a bit more than nervous. Fire onboard any fabric-covered airplane is enough to make anyone more than a little nervous.

Once down, I had to extinguish the fire, started by hot exhaust gasses escaping from a small hole that had become larger over time in the collector ring of my R-1340 engine. Those gasses had ignited the oil-soaked webbing that isolated the oil tank from the steel straps that held it in place. The webbing had burst into flames. Once I was down, I found I could not position the fire extinguisher between the cowling ring and the firewall; I thought for a moment I would have to abandon my airplane. Paskwachi Lake was, as I have mentioned, about two miles wide. I had to get control of that fire or swim a mile to shore. As a boy, I had managed to swim over a mile, but I was not in top swimming shape now and did not wish to tempt fate.

The fire was spreading as I climbed on top of the engine cowling, quickly loosened the Dzus fasteners. I was able to remove the cowling, and the extinguisher could now be used to quench the flames. In moments the fire was out. I used the high frequency radio again. "Lynn Lake, DRC, I'm down on Paskwachi Lake ... I have the fire out and I'll drift to the south shore. I'd appreciate a pick-up if that can be arranged."

As we drifted toward the south shore of Paskwachi Lake, I was amused to recall how I'd always used the Paskwachi River as one of my checkpoints when heading for Jackpine Island to pick up fish loads. The young Cree lad who usually operated the radio would always refer to me as "the Paskwachi kid."

Now, both DRD and DRC were out of action, and DFD had propeller problems; our only other aircraft was Aeronca Champion, CF-IRR, which would not be of much help in this particular situation. "DRC, Lynn Lake, I'll try to get Paul Ricky to come out for you with his Norseman," I was told. Paul brought Lloyd Smyth, our engineer, and his helper, George. In short order they had diagnosed the collector ring problem and we were on our way back to Eldon Lake.

Later that day, I flew Lloyd "Smitty" Smyth to Hughes Lake using Aeronca CF-IRR; Lloyd helped the rest of the crew remove the collector ring from DRD, after which Paul flew us to Paskwachi Lake together with that essential hardware. We left Smitty and George to install DRD's collector ring on DRC; Paul and I flew on with my load to Ingvar Stolberg's camp and we picked up the load of fish I was supposed to have picked up in DRC.

On July 31, just three days after the fire, I was again able to take DRC out of Paskwachi to carry on with my normal routine. We needed several trips into Hughes Lake before DRD was ready to fly; I have photographs of the engine change done on DRD in the wilderness on Hughes Lake. On August 10, I was finally able to get good old DRD back in the air, and, after making a local test flight at Hughes Lake, 1 was again able to go on one more fish haul from Ingvar Stolberg's Jackpine Island. Although competitors of sorts, Paul did not hesitate to come to my aid; such was the code of bush flying in those days. No pilot worth his salt would refuse to help another pilot who was in trouble.

About six weeks after the blown jug on DRD and the fire onboard DRC, during which time Paul Ricky came to my aid and transported me and my undelivered cargo, Taylor Airways was given the opportunity to repay the kindness.

As a commercial fisherman, Ingvar Stolberg had many dealings with fish inspector Rod Lynn. Ingvar had come to town on one of his rare visits, and a number of us gathered at Eddie Dyck's place one evening; John Foster was invited, as was our engineer's helper George. Ingvar Stolberg and Rod Lynn were there as well. TransAir radio technician Ben Telko was visiting Louise and me. Due to the fact we shared the house with Eddie Dyck, we asked Ben if he would join us at the party.

Everyone seemed to be getting along well, and there was plenty of beer on the kitchen table. Louise and I have never been beer drinkers. We were sitting on a chesterfield that backed against a windowsill on which an earthenware pot containing a fairly large geraniulm was perched. Ben Telko was sitting between us, and I believe he was in the process of downing the remains of his only beer of the night.

Suddenly and for no apparent reason that we could ascertain, I heard Ingvar shout, very loudly, "You're a liar!" It was directed at Rod Lynn. Who in turn yelled back at Stolberg, "You're nothing but a fat pig!" Ingvar Stolberg was a fairly heavyset man, perhaps sixty pounds heavier than he should have been. Rod Lynn at least had the fat part of his insult correct. These two staggered to their feet and were about to fight. Louise leapt to her feet and was about to step between them; I jumped behind Ingvar's chair and grabbed his shoulders, pushing him back to a seated position. At that instant, Rod Lynn threw a hard punch, missing Ingvar but catching me on the chin. I flew backward onto the chesterfield, and the back of my head smashed the geranium pot into a thousand pieces, with dirt and geranium stems flying in every direction.

I bounded off that chesterfield and hit Rod Lynn with my right; I could see his knees buckle. As Rod went down, I followed up with two more punches, using nearly all my strength. Rod Lynn lay on the floor, with tears running down his cheeks. "Why did you hit me, Al?" he asked.

"Because you hit me first, Rod," I told him. When we looked around, Ben Telko was nowhere to be seen. In fact, we have not seen him to this day. Ben was a nice, quiet fellow who hated trouble of any sort. The party broke up very quickly after that.

The crew told me that I had sealed my doom. Rod Lynn went on his fish inspections during winter using a dog team; in summer he used a canoe. Obviously he was in superb physical condition. "Nobody knocks Rod Lynn down without paying the price, and the next time you see him, he will get his payment from you," I was told.

I was aware of Lynn's physical prowess, but I felt I was fully justified in my actions. He might be able to take me, but at the same time, I was six-foot-two and 225 pounds. I loaded and unloaded diamond drill rods and 1,800-pound loads of whitefish from the Norseman. The very next morning I was in the town of Lynn Lake and walked past the cafe. I could see Lynn sitting at a table inside; I felt that if there was to be a problem between us it would be to my advantage to pick the time and place. I threw open the door and strode directly toward his table. My entry seemed to unnerve Rod Lynn. When I was within ten feet of his table he said, "G'mornin' Al, d'you want some coffee?"

No fisticuffs, no recriminations. Not even a comment on the previous evening.

The first trip I had made when I started with Parsons Airways Northern Ltd. had been to Dewdney Island, on Reindeer Lake. It had been an inspection trip by some prospective buyers who intended to establish a tourist fishing and recreation lodge there. It was a long flight, about 310 miles return, especially for a young pilot unfamiliar with the country over which he flew.

Later on, the developers built Arctic Lodge on the larger island. During the previous winter, they had taken a D-8 cat tractor to a nearby island and built a landing strip suitable for Douglas DC-3s. The idea was to fly their tourist customers from Minneapolis directly to the lodge, thus bypassing the inconvenience of changing airplanes at a stopover in Winnipeg.

One suspects that there must have been political clout to make this possible, since TransAir operated a daily service out of Winnipeg to Lynn Lake (except Sunday). TransAir, in turn, protested this operation, on the grounds that it produced unfair competition and loss of revenue that was justifiably theirs. It took some time, but finally Ottawa decided that TransAir had a good argument. An order was issued to prevent further flights into Dewdney Island. The cease-and-desist order was to be effective immediately.

I cannot now remember why TransAir was unable to make the pick-up using their own Norseman, but in any event, I was dispatched to bring guests from Arctic Lodge to Lynn Lake. TransAir would fly them to Winnipeg. Their Minneapolis DC-3 would meet them in Winnipeg for their return to the United States.

One of the guests at the lodge was Milton Eisenhower, and when I arrived to execute the transfer, Eisenhower protested. Arctic Lodge, as were so many other camps, was equipped with the Manitoba Telephone System two-way radio telephone terminal, which was mounted on a shelf suspended between two trees, over which was a roof to protect a user from rain.

I explained to Milton Eisenhower that TransAir felt the direct flights were unfair to their operation and that Ottawa had agreed. The result was that it was necessary that I transport all guests to Lynn Lake, where TransAir's DC-3 service would fly them to meet their own DC-3 in Winnipeg.

Eisenhower's brother Ike was president of the United States at the time. Milton placed a call to the White House through the MTS installation at Arctic Lodge. Normally, incoming signals were reproduced by loudspeaker. This was only for incoming calls, but when the telephone was being used, incoming voice signals were routed to the telephone handset in order to preserve some semblance of privacy—of course, being a two-way radio circuit, everyone at the other camps could hear.

I heard Mr. Eisenhower's one-sided conversation. "Hello Ike, Milton here," he said, "Listen Ike ... the Canadian government has canceled our flight. We are stuck up here in Canada without any way of getting back."

This was certainly not true.

"Well, yes ... we need that flight in here today." He was silent for a few seconds and then. "Okay Ike ... you'll get back to me, okay?"

"Well ... you can call Manitoba Telephone System at The Pas. They can get me here at Arctic Lodge." Another second of silence.

"Okay Ike, thanks ... g'bye."

In a short while, the MTS operator at The Pas called and Milton was back on the telephone handset. He said very little this time and I couldn't hear whoever was talking to him. When he hung up, he turned to me.

"I'm sorry," he said, "but we will not be riding with you today because Ike has talked with Prime Minister Diefenbaker and the ban on further flights has been lifted for this occasion only. I am sorry that you have made this trip for nothing. I hope we have no hard feelings."

I was then and still am disgusted that dirty politics extend to such trivial matters. Milton Eisenhower gave his brother the impression of being stuck with no escape unless big brother could work a miracle on his behalf. Ike had worked the miracle.

The fact that John Diefenbaker played along still makes me angry. He could and should have simply stated that this was Canadian law, which he had no intention of breaking. Power corrupts. I believe the same kind of U.S. pressure was used for the cancellation of Canada's Avro Arrow project. It was in that same general time frame, and the same people were involved.

The American DC-3 made one final flight into the strip near the Arctic Lodge campsite on Dewdney Island. I had to return to Lynn Lake with an empty airplane and 150 miles of non-revenue flying behind me in Norseman CF-DRD.

The summer season was quickly winding down, but there was still more fish to be picked up at Jackpine Island. Ingvar Stolberg was on Reindeer Lake, as I have mentioned. The lake occupies a goodly part of northern Saskatchewan, while its northeastern portion bleeds into Manitoba.

Ingvar and I had a good business relationship for my entire Lynn Lake stay. Most all of my dealings with him had been congenial, but there was one exception. That day he complained that I might not make a last flight to pick up his remaining fish. It was getting late in the day, and he complained that there wasn't time left for me to pick up the last load.

"My business is very important to you; I've got fish that will spoil if you don't come to pick it up," he told me.

"Thanks a lot, Ingvar," I said. "I've been rushing like a mad fool all day just so I could pick up all the fish; I haven't eaten anything since breakfast, and every damn fisherman in this country says his fish will spoil if I don't pick it up, as though their fish was the only fish that will spoil. You're no special case ... I told you I'd pick up your last load, and I damn well will. Thanks a lot for your vote of confidence." With that, I scrambled into DRD's cockpit and blasted off directly from his dock, making a good deal of noise and leaving it in my wake. My reaction was, in part, I suppose due to my run-in with the fish inspector who had thrown a punch at Ingvar but landed it on my chin.

Upon my return, Ingvar met me with a beautifully prepared sandwich and a steaming cup of hot coffee. No mention was ever made of Ingvar's complaint or my show of anger. From that day forward, we had no further complaints from Ingvar, nor was there ever any show of anger on my part. We remained the best of friends. For our son's first birthday, Ingvar gave Mark a gift of cash.

On September 30, 1959, I had my final pick-up of fish from Ingvar's Jackpine camp. As I had done nearly seventy-five times before, I called him on the radio. "Jackpine Island, DRD, by Paskwachi River on the hour, 3,000 feet ... estimating Jackpine 16:25 Zebra."

"Roger," came the reply, "We vill have this load ready fer you when you get here."

The weather was getting colder and storm clouds were building on the western horizon, with diminishing visibility. I began to wonder if I would make Jackpine Island, let alone make the trip back to Eldon Lake.

"Jackpine, DRD," I said, "You had better put a hold on the load for now, Ingvar ... keep it on ice. This weather seems bent on putting a stop to our flying operations for a while."

"Okay," came the reply, "We'll just wait till you get here then."

"Okay ... DRD," I said.

The weather closed down just after my arrival and for the next two days, while we waited for a break, temperatures dropped lower than we had seen during the entire season.

Finally, on October 2, we awakened to good visibility but even lower temperatures than the previous night. The fish were removed from the icehouse and loaded on board DRD, 100 pounds per washtub, and eighteen tubs for the 1,800-pound load for my flight back to Eldon Lake.

When 1 arrived at our base, I found the entire lake frozen over and I was faced with landing a seaplane on ice. Since it was my first such landing, I didn't know what to expect.

I had no choice, so I lined up on what I hoped might be a near-normal landing. I was cautious as I gingerly approached the ice surface at about ninety miles per hour, carrying plenty of power to descend at less than 100 feet per minute. After a fairly long and very flat trajectory, I was gratified to hear the keels of my floats come into contact with that icy surface. What I had feared might end up with my losing control became a normal landing under complete control; when my forward speed had dropped to about five miles per hour, the surface began to sink at a leisurely pace; large flat sections of ice tilted out to the sides as DRD settled into water. Progress to shore mimicked the action of an ice breaker: my floats would climb onto the edge of the ice surface, then DRD's weight would force the ice to submerge and shatter, and millions of ice shards were left to float in my wake, sparkling like myriad diamonds under a pale sun in that cold October air.

When we reached shore, I used full takeoff power to move CF-DRD up the plank ramp which we normally used for maintenance. Parking DRD just short of the A-frame (used for engine changes) and away from the ice at her heels, I shut her down.

It would be thirty-three years before I was to see DRD again. On October 9, one day after Mark's birthday, Louise and I boarded a TransAir DC-3 at the Lynn Lake Airport, all bundled up in our winter wear. We said goodbye to Jack Deacon, who was TransAir's airport radio operator/agent. Jack had been the Lynn Lake agent following my gall bladder surgery. I had come to know him well. We departed Lynn Lake and headed for Winnipeg to be greeted with warm sunny skies. The green grass in Assiniboine Park appeared almost like paradise, as if winter had not even crossed Mother Nature's mind. Unfortunately, I had not thought to snap a picture of DRD as she rested on that ramp following our final flight and ice landing. Somehow, it never occurred to me that I might not return to fly her again. Perhaps I just didn't want to contemplate the possibility.

Bush and Arctic Pilot

A. R. (Al) Williams

18 Back to the Future

The next time I stood by DRD's side was when I visited Red Lake for the opening of Norseman Park in July 1992. Wearing a fresh coat of paint, just as I remembered her from thirty-three years earlier, she rested serenely on her gracefully fashioned pedestal.

The story of my reunion has its beginnings late in the float season in 1959. An itinerant aircraft stopped at our Eldon Lake base and took on aviation 80/87 gasoline. I had come in from a fish haul; I had taken the logbooks in and placed them on our counter. As I was ready to leave on another trip, I could not find the journey log; engine and propeller logs were there on one end of that countertop but the aircraft journey log was gone. It didn't occur to us that the itinerant aircraft had departed with our logbook, unaware they had done so.

Red Lake is considered "The Norseman Capital of the World" because as many as sixty-eight Norseman aircraft operated there at one time. The people of Red Lake decided to capitalize on the title. In keeping with these traditions, Red Lake decided to build "Norseman Park" right against the north shore of Howey Bay, just below their main street and overlooking the bay.

Volunteers cleaned up the debris of rusty cans and broken beer bottles from the shoreline immediately in front of the Canadian Imperial Bank of Commerce. They hauled in truckloads of gravel and topped off this base with dirt and sand, built multitiered terraces, complete with landscaping and gravel paths. Visitors can now stroll the paths to read historical captions that complement the weatherproof photographs of the Red Lake flying history stand.

Near the west end of this park, a huge concrete and tile circle has been constructed, fifty-eight feet in diameter, over which a restored Mark VI Norseman is suspended on a beautifully curved cantilever twin-boom pedestal arc.

The Norseman, which wears its original bright yellow paint with a red speed-line and wingtips, is my beloved CF-DRD. When I heard she was to be placed on that pedestal at Red Lake, 1 simply had to go see her.

I flew out of the Edmonton Municipal Airport (it will always be the Industrial Airport to me) on board a Fairchild F-28. After a short stop in Saskatoon, I arrived in Winnipeg around 11:00 hours. I'd not been near Winnipeg since October 1959, and hardly recognized a once very familiar airport.

It had been my hope that Robert Noorduyn Jr. (son of the designer of my favorite aircraft, the Noorduyn Norseman) might attend the opening of Norseman Park. All the thousands of hours I spent in the left seat of various Norseman instilled a love and trust in this reliable old aircraft. In Winnipeg, to reach Red Lake, I changed airplanes from an F-28 to a Beech 1900-C via Bearskin Airways.

I turned to the mature man sitting directly across the aisle in the Beech and said in a friendly manner, "Hello, my name is Al Williams." The gentleman extended his right hand and smiled. "Bob Noorduyn," he said. Bob and I became close friends during that flight from Winnipeg to Red Lake, and we exchange correspondence to this day.

I had also hoped to meet Robert S. Grant, pilot and author, well known in Canadian aviation circles, but I didn't know anyone else in Red Lake. Bob Grant flew a Canadaire CL-215 water bomber for Ministry of Natural Resources of Ontario. I had been district radio operator for that ministry when it was called the Department of Lands and Forests. I had enjoyed Grant's article about Red Lake and the Norseman in *Canadian Geographic* magazine.

I was therefore surprised when I became an instant celebrity in Red Lake. An employee of Buffalo Airways at Yellowknife was looking for something in the attic of their office one day. He happened upon a logbook for CF-DRD. He had no idea of what aircraft it was or how the book had come to be in their attic. He asked around, but nobody could ever recall such an aircraft having even been in that part of the country.

He put it out of his mind until the Norseman Park opening advertisement appeared in *Canadian Aviation*. He went to the attic and brought down the logbook. Company president Joe McBryan thought the logbook might be the only such one in existence. He decided that they should fly the log book to Red Lake, to be placed in the Norseman aircraft museum. Joe was correct. DRD had been a derelict heap for several years; it was missing its engine, prop and journey logs. My recovered logbook had disappeared from Eldon Lake back in 1959 during my days at Lynn Lake.

I arrived at Red Lake for the park opening about 14:30 hours that day in 1992; Joe and his men had landed their Cessna 185 around 15:15 hours. I was having a Norseman burger in the Lake View Cafe ("Pump yer floats before ya eat this one" the menu proclaimed). "Are you Al Williams?" I was asked by a local resident. "You flew DRD, did you?" asked the stranger. I was flabbergasted. How would anyone, outside of Bob Grant, know that I had flown DRD?

The answer was the logbook flown here from Yellowknife. Of all the journey logs which must have been completed on DRD over the years, this one was filled, line after line, page after page with my own signature from my days at Lynn Lake in 1959. Thirty-three years had passed by, but those years all came tumbling back to me in a rush when I read my old log. Each entry brought back an instant replay of a particular trip.

The day Milton Eisenhower refused to return to Lynn Lake via DRD and whined to his brother Ike, the president, until John Diefenbaker allowed just one more DC-3 flight to Arctic Lodge; the day I blew a jug near South Indian Lake; that trip with Murdo Clee and his offer to pay any court fine imposed on me—now, those flight entries were available for any visitor to the Red Lake aviation park to read.

As for my visit, it was amazing that so much was crammed into those three days. A very young Scott McAllister (about twenty-two years old) gave me a ride in Norseman CF-OBE. OBE was the only Norseman to be metallized—wings and all. It was interesting how the years disappeared as I again experienced the smells, sights and sounds of flying a Norseman. I had a lump in my throat as I stood there beside DRD remembering all the wonderful adventures we had shared in days gone by.

19 Down to Earth

After I fulfilled my promise of a one-year contract to Jim Shore at Taylor Airways, I began to consider other employment opportunities. I had missed out on my opportunity to fly with Pacific Western Air Lines, although it had been my own decision to finish out my time with Jim. But the idea of my flying with a large airline began to gnaw at my mind. In response, I applied for a pilot position with Trans-Canada Air Lines (TCA), now known as Air Canada.

I received a very encouraging letter from Rene Giguere, who was then Trans-Canada Air Lines' chief pilot. Mr. Giguere advised that while TCA had a policy of not hiring pilots over the age of twenty-five (I was a few months beyond that age) he was impressed with my flying experience, and he felt I would be an asset to Trans-Canada Air Lines.

In a few days I received a copy of Rene Giguere's letter to his airline in which he did his best to secure a position for me. Head office didn't make any concessions with regard to the age of an applicant. Three years later, the company increased their maximum age to twenty-eight years. However, I was a few months over their new age limit and was again rejected.

In casual contact with an amateur radio station in Calgary a few days after my initial rejection by TCA in February 1960, I was told about Electronics Service Supply Company. The company was in need of a technician with a commercial pilot license. It sounded almost too good to be true. I called the manager at Electronics Service Supply in Calgary, Lorne Lebon, and we set up an appointment for the following Wednesday. I drove to Calgary from Lac Vert in Saskatchewan for an interview.

When I arrived in Calgary, I discovered that the contract on which they had been negotiating had been canceled. They had no job for me now. I would have flown to various sites, servicing electronic equipment. I was to have been given a late-model Cessna 180; while my pay would not have equaled the salary of both technician and pilot, it would have been substantial. The job was very attractive for me, and the position was cost-effective for Electronics Service Supply.

Lorne felt badly that the job had fallen through, and he set up a meeting for me with Benny Redisky. Mr. Redisky owned some television cable systems in the East Kootenay area of B.C. Redisky was looking for someone to service his television system in Fernie. Redisky and Jim Gillespie, who flew their Cessna 180, were arriving in Calgary that afternoon; a meeting was set up for the following morning. Gillespie was partner in the cable business and also operated a flying club in Cranbrook.

Benny Redisky was well known as "Benny the Burglar" to anyone who did business with him. Unfortunately I knew nothing of Redinsky's unsavory business reputation. I had faith in Jim Gillespie because he was a good friend of my friend, Roy MacMillan, in Sioux Lookout. Roy and I met as operators of amateur radio. I reasoned that a friend of Roy's had to be a good and honest man and that they must surely run an honest operation.

The job they offered me would mean a move to Fernie to set up a radio and television servicing shop. Redisky promised a $250 per month retainer for service to his cable system. At that time, $250 per month was enough to provide a living even without an income from the radio and television shop. On February 19, 1960, my wife Louise, her sister Charlotte and her husband Earl Carefoot arrived with me in Fernie. During the night we had a very heavy thunderstorm, which seemed incredible to us, as we had never experienced a winter thunderstorm before.

We had come to Fernie to examine the possibility of establishing an electronics servicing business, combined with the promise of a cable service contract from Benny Redisky. We quizzed the locals on opening a radio and television service shop and were assured such a facility was needed. I decided to establish Rocky Mountain Electronics. I was able to raise only $1,000 operating capital, which is certainly not enough to start any business. As I was not aware of this business fact, I went ahead and did it anyway.

Fernie was and still is a coal-mining town. Money was scarce. I had a fair amount of electronic service business, but there were a good many folks who would ask if I could carry them until the next paycheck. There were times when I had to go out and collect outstanding accounts on a Saturday morning, so Louise could go and buy our groceries in the afternoon.

Benny Redisky never kept his word to me with regard to the contract to service his cable system. By the time I was aware of his sleazy maneuver, I was stuck in a business that brought in enough money for living expenses but very little else.

When Crow's Nest Pass Coal Company began to find a need for two-way radios, I was able to establish sales and servicing of the equipment as part of my business. We never went into the red, but one day our bank account dipped to fifty-six cents. The electronic repair business certainly did not turn out to be as financially rewarding as flying; we barely kept our heads above water. However, I was home each night. By June 1, the emotional need to fly became overwhelming. Although money was scarce, we went to Cranbrook, where Jim Gillespie ran his flying school. I soon discovered that 1 had been wrong in my judgment that anyone who was a friend of Roy MacMillan must be a good and honest man.

There I was, a commercial bush and Arctic pilot, with thousands of hours under every conceivable flying condition. My last flying had been done in October, and now Jim Gillespie wanted to ride with me on rental-plus-instructor charge for forty-five minutes. Any pilot worth his salt knows, after one circuit, if the pilot being checked out is safe. Jim Gillespie just wanted to take advantage of a situation.

After one circuit, I landed and taxied to his ramp. He said he wanted me to do several more circuits.

"Jim," I said, "do you want to rent this airplane or not?" He replied he did want to rent it. "Well, Jim," I said, "I either take it now, or let's just forget it. I have no intention of paying for you to be a passenger." With that, Jim Gillespie removed himself from Cessna 150 CF-LXJ and I took Mark for a sightseeing flight.

Finances had picked up some by June 24, 1961, and we had gone to Calgary to purchase more electronic parts for our business. Naturally, I wanted to fly again. I went to the Calgary airport and rented Cessna 175 CF-LBF for an hour and a half of sightseeing in the area.

On April 22, 1962, I flew a Cessna 150 CF-MOR out of the Medicine Hat Municipal Airport, but for only twenty minutes, just so I could feel the joy of flight once again. My two-way radio business was picking up a bit due to the Department of Transport's ruling that made it much easier to acquire citizens' band radios. I sold and serviced Johnson messenger radios. Crow's Nest Pass Coal Company was my primary customer, although I did travel to southern Alberta to install some equipment at a few ranches in the foothills north of Lundbreck Falls.

Time marched on, and our twins Kevin and Julie arrived on January 21, 1961. Business continued much as it had since the beginning; we were able to take the odd trip back to Medicine Hat or Saskatchewan, but there was very little surplus money. We were able to move out of our thirty-five-dollar apartment (a bargain even then) and buy our first house. This was a marvelous turn-of-the-century home with a four-foot-wide front door with a door knocker shaped like a devil's head. It also boasted a wide entrance and broad stairway leading to the upper floor. We enjoyed the huge fireplace, but a day came when we felt a bit less than joyful about the house. Our sewer pipes backed up, and I simply couldn't afford to hire a plumber. I did a great deal of back-breaking digging in our back yard while George Quail, the local plumber, supplied new sewer pipe at cost together with plenty of much-needed free advice.

After seven years of hand-to-mouth operations by my electronics company, the Crows Nest Pass Coal Company miners went out on strike. While I continued to work, I found most of the local folks were unable to produce enough money to pay entertainment bills. They were good people, but there simply was no money left to pay for repairs to their electronic home entertainment equipment. I didn't have the heart to turn them down, even though I was just as badly off financially as they were. To their credit, however, as things began to turn around, these people came to pay their debts to my wife and extended their thanks for credit. I was no longer in Fernie, however.

In order to feed and clothe my family, I was touring the country looking for flying work. One application I sent was to Gord Bartch (who had been co-pilot with Bob Lundberg when CF-CAR had quit for lack of fuel). Gord operated an air service with its head office in Calgary. I didn't get that job, quite possibly, because of my flippant remark when Gord told me he had in mind a job flying his company's Found FBA-2C out of Norman Wells. "It would be hard to get lost in a Found," I joked. I am certain that no prospective pilot employer wants to hear the word "lost" even in jest.

I applied for a position with Keir Air Transport, Edmonton. That company felt that my flying skills were not "current" and therefore turned me down, as did Gateway Aviation in Edmonton. I had flown only a limited number of hours during the past seven years. It is interesting, though, that the moment Bob Lundberg at Northward Aviation hired me, Gateway Aviation management called me at the motel where I was staying while I waited for Cessna 206 CF-MWV to be readied for my summer of flying in the Northwest Territories. I had to turn them down. About a year later, I had an opportunity to assist Mr. Keir with a technical problem that had no connection with aviation, and he said he had been foolish not to have hired me when he had the chance.

It is a sad fact of life that when you are unemployed and need a job, prospective employers won't hire you because they are leery of the fact that you are not working. On the other hand, when you are employed for a company and not in dire need of income, other companies are often anxious to hire you.

I would miss Fernie's mountainous beauty. Trinity Mountain (The Three Sisters) and the "Ghost Rider" silhouette that can be seen on Mount Hosmer. I would also miss the people with whom we had become friends: Bob and Bill Shenfield and Bob's sister, Nellie. Their friendship would be treasured always. Bill and Mable Mawer helped us a good deal, especially because Bill offered service advice to me as an amateur radio operator. I left my family in Fernie—Louise, Mark, Kevin, Julie—as I set out to fly in the Northwest Territories.

Bush and Arctic Pilot

A. R. (Al) Williams

20 Northward Aviation Adventures

I found myself installed in a hotel room at Fort Smith, just on the border of northeastern Alberta and the Northwest Territories.

Gordon Pachal was away in some southwestern area of the territories with de Havilland Beaver CF-HEP; he would be back to Fort Smith, his regular base, before long. Upon Pachal's return, we would meet at Hay River, where we would trade airplanes; Pachal would take back his faithful Cessna 206 CF- MWV and I would get that Beaver, for some unspecified length of time, at Hay River.

While I was at Fort Smith, things were very quiet and I was not able to secure more than one sightseeing flight for some tourists who wanted to see a relatively small portion of the country around Fort Smith.

I had flown MWV a grand total of 645 miles in two days, which included my ferry flight from Cooking Lake to Fort Smith, when I received a call that the Cessna 206 was required in Hay River. No word had yet come down as to when I would trade with Gordon Pachal.

On the ferry flight from Fort Smith to Hay River, I heard a familiar voice on VHF Radio at 126.7 MHz. Len Lavoie, one of my old instructors at the Winnipeg Flying Club, was now the captain on a commercial flight from Yellowknife to Churchill. We had a brief and friendly reunion via the radio. I continued on with my flight, exploring new sights and sounds.

Soon after that 364-mile flight, I found myself questioning the ethics of our fuel agent in Hay River, who was not an employee of the company. He was suspected of being less than honest with locals as well as out-of-towners. He ran a vehicle rental service, and there were some questions about his renting the same vehicle out to two or three different companies at the same time. I had my own suspicions about him padding the bill because I had gone from Cooking Lake to Fort Smith plus a local sightseeing flight (a distance of 645 miles) in five hours and ten minutes, at a power setting of 55 percent. Fuel consumption was just more than forty-six gallons—a bit over nine gallons per hour. Yet, when the agent had refueled MWV in Hay River after only 364 miles—just over three and a half hours—he billed for 46 gallons, which would mean a consumption of thirteen gallons per hour: my power setting had remained at 55 percent. The agent had padded my fuel bill by 15.25 gallons.

It didn't occur to me to question that fuel consumption until I was alerted to the agent's reputation. At that point I did some calculations and came to realize that I should have consumed only 30.8 gallons on the flight from Fort Smith to Hay River.

The locals refused to charter our Beaver unless there was no alternate aircraft available because of the agent's crooked accounting.

I wrote to Northward's general manager, Colin Grant, pointing out our dilemma. Colin replied with a note of apparently sincere thanks, but no changes were ever made while I was there.

On July 4, 1967, our chief engineer, Frank "Sok" Sokolowsky, and a helper wanted to go to Fort Smith. We climbed aboard CF-MWV, and I took off. There was very high humidity and the clouds were low as we departed one of the many river channels that make up the river delta at Hay River. Twenty feet above the trees, we encountered a few tendrils of cloud reaching into the treetops. Suddenly, moisture entering the air vents fogged our windshield so that we lost forward visibility almost completely; I had to land as quickly as possible, but the confines of those fairly narrow river channels would make landing a bit more tricky than I had anticipated.

We tried wiping the inside of the windshield, but almost at once it would fog over again. I had to make several passes over our intended landing path before I was satisfied that I would be able to put that Cessna 206 into the proper river channel without hitting any trees on the right side.

I lined up by sighting out the left window, staying as close as possible to the left bank, maintaining a margin of safety, knowing that if I could miss the trees on that side, our right wing would remain clear of trees on the other side. We all felt a lot better once our floats were firmly back on the water.

The following day, Gordon Pachal landed Beaver CF-HEP at Hay River. Before we exchanged airplanes, Gord, who didn't know me or my background, gave me a proper check ride on Beaver CF-HEP, as it was his responsibility to make sure I could handle it before releasing it to me.

I knew that the Cessna 206 was "his" airplane, and there was absolutely no need for me to check him out in her. I assumed the left seat in Beaver CF-HEP and took off out of the one river channel that ran directly past the Northward Aviation dock at Hay River. Gordon was apparently satisfied with my takeoff, but when we returned for our landing he became a bit nervous. Since I was very familiar with Beaver aircraft, I approached the channel a bit higher than Gord would have done; I pumped down full flap as I reduced throttle. Our descent was steeper than Gord was expecting, and I could see he was somewhat nervous; Gordon began retracting his feet toward his seat and dropped his hands to the sides of that seat in anticipation of a harsh arrival. He relaxed as my floats came into contact very gently with the surface, and we laughed a bit. He commented that I had obviously flown a Beaver in the past.

Later that same day, I had a flight to Yellowknife with Canadian National Railways (CNR) communications personnel. They were beginning a summer of the Mackenzie Valley upgrade, and many improvements were made to the existing telephone system which provided telephone and broadcast radio programming for the Northwest Territories. I did a good deal of flying for the CNR before summer ended, mainly out of Norman Wells.

On July 7, I was called upon to fly in support of firefighting around the Deschaine Lake area, just north of Buffalo Lake.

There are five or six small lakes shown on the map as a single lake named Deschaine Lake. I flew three hours and twenty minutes servicing that fire, hauling men, pumps and hoses into those lakes.

Because of our crooked agent, I didn't have much flying out of Hay River. There were a few trips, to check out a drill camp or fly a fisheries inspector out to a boat operating in Lonely Bay, about eighty miles north of Hay River.

On July 11, Bob Lundberg called me from Yellowknife; he asked me to catch a Pacific Western Air Lines Convair flight to Yellowknife. I had liked Bob Lundberg from the first time we met. He held the position of chief pilot when I flew for Northward Aviation in 1967. (He left that position for a stint in a twin Otter flying from Edmonton to Grand Cache for a while. Then he flew a DC-3 and later he flew 5,000 hours as captain on de Havilland Dash-7s—a record for anyone flying that model. Bob finally left commercial aviation at age sixty-five. He'd had too many pilot medicals—too many check rides.)

Bob said to leave my baggage in Hay River, as I would have no need for my things because I would be returning on the next flight. I soon found I had been misled. Bob had given the order in good faith, but things took on a life of their own. The pilot who was supposed to arrive failed to show, and it was necessary for me to fly another Beaver, CF-NOT, to Cambridge Bay on Victoria Island. My one regret was that I had left everything, including my earphones and microphone, in Hay River. When I returned, I found someone had stolen them from CF-HEP.

I'd been away from commercial flying for nearly nine years, and I hadn't flown even a private aircraft during the past five years; not until June 28 (just three weeks earlier). As I laid out my maps on the hotel floor that night, I certainly had some doubts about my rusty abilities. *I'm just a Saskatchewan farm boy*, I thought to myself *What am I doing here, preparing to navigate into the Arctic?* I pondered that question for some time.

Wait a minute, I thought. *You are a Saskatchewan farm boy no longer; you are a professional bush and Arctic pilot.*

It was true I had navigated my way to Arctic Bay, on the north end of Baffin Island, and returned to Churchill. I suddenly remembered that Arctic Bay was quite a lot further north than Cambridge Bay. Cambridge Bay is only 69 degrees 19 minutes 26 seconds north and 133 degrees 35 minutes west, while Arctic Bay is at 73 degrees 17 seconds north and 85 degrees 1 minute 59 seconds west, nearly 300 miles further north than Cambridge Bay.

After waiting impatiently for my aircraft to be refitted with floats and outfitted for summer, Bob and I flew out to Cooking Lake. Bob took the left seat and we taxied out to the runway atop a wheeled carriage, which is usually referred to as a dolly. Bob always referred to our engine run-ups at the end of the runway as "getting up nerve."

Cessna 206 CF-MWV sat in a special cradle, with her floats cushioned by felt. A light line was tied between the rear spreader bar of our floats and the brake lever on the dolly. It turned out, neither of us had ever made a takeoff from such a device; the result was that neither of us really knew what to expect.

"I'm going to run her to well above her stall speed," Bob said. "I'll tell you when, and you hit the flaps same time I rotate, all right?"

"Okay," I said, and Bob opened up to full takeoff power.

At about sixty-five miles per hour, Bob called "Now!" I hit the flaps as he rotated. Our Cessna 206 fairly leapt into the air, and the line parted as it moved the brake lever. Looking back through the rear window, I could see our dolly come to a straight ahead stop; the brake had certainly worked well enough. The runway seemed smaller with each passing second as we rapidly climbed away.

Our flight from the Edmonton Industrial Airport to Cooking Lake was uneventful. But it was great to be back in the air once again. Now, thanks to Bob Lundberg, I was to have a modern airplane. I had sorely missed flying during my years of being earthbound.

Not long after I had checked out—the same day, as a matter of fact—I was on my way to Fort Smith. Just before my departure, Bob said, "Until you know your airplane better, stay out of the snye."

I had to ask what a *snye* was. Fort McMurray, a site en route to Fort Smith, boasts a snye, a backwater that flows into the river but remains calm water, relatively undisturbed by the flow of the river. Fort McMurray's snye has been used by seaplanes ever since float-equipped aircraft have operated there.

I decided Bob was right; I would avoid the snye until I had this airplane and its idiosyncrasies mastered. I maintained my resolve even when I noticed my battery power had gone. First, I noticed the radio getting quieter. When it died completely, I realized my battery was dead.

Many ammeters in aircraft show both charge and discharge, but the one in MWV's panel began at zero and indicated charging only. Since I had no battery power, I would continue to Fort Smith and worry about the problem when I arrived. Ignition was not a concern, since airplanes use magnetos for that purpose.

The seaplane docks at Fort Smith occupied the south bank of the river, downstream from a dam known as "rapids of the drowned." This dam occupies the entire width of the river. Lundberg had told me our northward dock occupied the position nearest to the falls and that there was a back-current swirl in the river below the dam, which made for difficult docking. Now it came to me—if I missed the dock on my first go, the swirl would carry me into the rocks beyond, and I'd be unable to restart my engine.

The gods smiled on me that day. My first docking attempt was successful.

Within a few days of exchanging Cessna MWV for Beaver HEP, Bob Lundberg asked me to fly to Yellowknife for a check-ride in de Havilland Otter CF-CZP. When that fully loaded seaplane version of Otter had been mastered, Bob said, "Let's go to the airport and I will get you a wheel endorsement." Of course I had no need of any wheel endorsement, but I was to have the pleasure of flying Otter CF-VVY on wheels. It was a real joy to do that, because I had never had the opportunity to do so until that day. My boss slid into the left seat on the flight deck and told me he wanted to demonstrate just what the Otter was capable of doing on wheels.

Bob stood on the brakes and opened the throttle to full takeoff power, then he released the brakes. We almost catapulted into the air as he leaned over into a fairly steep left turn, the port wing tip just a few feet from the ground. We came around 360 degrees and he put her wheels back almost exactly in our original tracks. Our flight had lasted maybe twenty-five seconds. Bob cautioned me that such a maneuver was not recommended unless conditions were ideal, as they were at the time (almost zero wind). He also insisted that a good deal of experience was needed on the aircraft to execute the maneuver. Then it was my turn to sit in the pilot's seat. Although we never discussed the subject further, I gather that Bob granted me my wheel endorsement.

Bill Mills had come on staff a few days prior to my arrival in Edmonton. We had become fairly good friends while we waited for our respective aircraft to be readied for flight. Bill had flown in Antarctica a year earlier; he had been hired to fly Beech 18 CF-SRK out of Cambridge Bay.

On July 15, 1967, I flew Beaver CF-NOT to Cambridge; I could fly the Beaver that distance because of its wingtip tanks. Most Beavers with standard tanks could fly only about 400 miles, but with that extra fuel in CF-NOT, reaching Cambridge Bay, a distance of nearly 600 miles from Yellowknife, was no problem.

It is interesting that when a pilot is map-reading unfamiliar territory that one often sees very little to either side of the line he has drawn on his map. And so it was on my first trip to Cambridge Bay in my new mount, Beaver CF-NOT. I wondered, later, why I had not noticed several fairly large lakes which would have been great landmarks, had I but taken notice of them. I missed seeing Duncan Lake, which lay almost at my feet; Gordon Lake was just off to the right of my flight path and Prang Lake was directly in my path. But I saw none of them; I followed the line I had drawn the night before.

I approached Contwoyto Lake, which is near the midway point from Yellowknife to Cambridge Bay. The radio installation at Contwoyto Lake was jointly owned and operated by the Department of Transport and Pacific Western Air Lines as a navigation aid, making any weather information available to aircraft passing by. The three men who manned the isolated station were always hungry to talk with a passing aircraft. Although most flights simply exchanged position reports and weather information, I used to have a little talk with the operator on duty. I felt that these jawing sessions of ours served as a good tonic for the men at the station.

My Beaver CF-NOT was equipped with dual Automatic Direction Finding (ADF) gear, and it was a great comfort to track outbound, with one ADF tuned to Yellowknife while the other was tuned to the Contwoyto Lake low-power nondirectional beacon (NDB). It had a range of something on the order of thirty miles and produced a circle of sixty miles diameter centred on the station. As long as one could stay within that radius, the Contwoyto Beacon could be used to fly directly to the lake. Normally, our navigation would get our flights within five to ten miles of that beacon, even in bad weather, and then we could track outbound from Contwoyto onto Cambridge Bay with the second ADF on Cambridge Bay. The 2,000-watt Cambridge Bay beacon radiating its signal from its 600-foot tower was normally useful several hundred miles out. There were times when it could actually be heard as far away as Calgary.

When I arrived at Cambridge Bay, the first seaplane of the year to arrive, our engineer's helper asked if I had eaten. I had not eaten since early morning, and he asked if I would like a steak. I expected a steak dinner with potatoes and perhaps a vegetable, but he presented me with exactly that, a steak—all by itself. It was perhaps the most satisfying dinner I'd had in some time. No expense was ever spared in providing the very best quality of food, not only by Northward Aviation but by all the other companies in the north as well.

Also when I arrived at Cambridge Bay, I was informed that Bill Mills had "come to grief" with his airplane on a little gravel airstrip at Spence Bay. Bob Lundberg was at the Cambridge Bay Airport. I was sent there to see if I was needed. Bob told me that Bill had put SRK up on her nose in the early stages of takeoff. "I've got to get out there and take a readin' on the boy," he said firmly. I took this to indicate some question about Bill's handling of what was assumed to be a routine takeoff. It turned out that the strip had a low spot not far from where Bill had begun his takeoff run. Recent rain had turned the soft spot even spongier. Bill hadn't gained enough speed when his wheels rolled into that soft ground. CF-SRK planted her nose in the gravel surface. Both propellers had been destroyed and the nose was damaged. Bob Lundberg loaded his second Beech 18 CF-RUA with two new props, an engineer, a box of tools and a few paying passengers. Away he went to rescue the downed CF-SRK and to "take a readin'" on Bill Mills.

Bob flew his Beech 18 on the approach to Spence Bay, but a wind had sprung up during their flight. To appreciate this situation, you have to realize that this strip is near a steeply rising hill on one side, and on the other side, the ground falls off toward the Arctic Ocean. The strip is narrow, and rocks at both ends make approaches tricky at the best of times. Just short of the approach end of the strip, a particularly violent wind gust tugged away at the Beech 18 and threatened to drive RUA into the hill. Lundberg applied a good deal of left rudder in order to avoid slamming into the hill. Just as the RUA was coming over the end of the strip, the wind forced her down so that her rudders snagged the rocks. Heavy left rudder deflection was already applied and the rocks crushed the twin rudders; they became locked in nearly the full left-rudder position. No amount of force could move them back to a neutral position.

At this point, the main undercarriage touched the strip but the left rudder was steering RUA to the left and the Arctic Ocean. Bob pulled the right throttle to idle and applied full throttle to his port engine in an attempt to straighten the aircraft on the runway. His thoughts came in waves: "Gee, I hope I don't do too much damage to my airplane." "Gosh, I sure hope we don't get hurt too badly." "Man. I sure hope we don't get killed!"

Finally, his left undercarriage hit a large rock out near the edge of the shoulder of the strip. It was sheared off under the engine nacelle. The left propeller then hit the ground and stopped moving, while the engine tried to keep it turning. CF-RUA came to a grinding halt. A very large man sitting in the right-hand seat, a man who would not normally be able to squeeze through the overhead hatch, went through that hatch and took the door with him. Before Bob could move, the man was standing out on the ground yelling for Lundberg to get out.

In due course, the two new propellers were installed on SRK and the nose was patched with strands of tape; SRK was flown out to Cambridge Bay. When they were all back, safe and sound, I sidled up to Bob Lundberg as he stood smoking his pipe and said, "Do you suppose somebody should 'take a readin' on our chief pilot?" Bob demonstrated his innate sense of humor with a slight grin. Remarkable, considering his recent experience.

In February 1961, Bob Lundberg and Gord Bartch, in DC-3 CF-CAR, were engaged in another Arctic adventure. They were headed out to a drill site belonging to Bawden Drilling, 250 miles west of Resolute on Cornwallis Island. The night was cold, and darkness flooded the land. A front was moving in from the northwest, and the radio informed that their destination was closing down due to low clouds and poor visibility. Their immediate reaction was to return to Resolute, but a radio check soon told them that it, too, was closing down. Checking with all other available landing strips confirmed that their weather was also deteriorating rapidly. They were stranded!

As time passed, the Byron Bay radar site on the Distant Early Warning (DEW) line advised them of clear skies and visible stars. A quick check of their charts told them Byron Bay was about 1,000 nautical miles south of their present position! Would the fuel stretch that distance? Lundberg called for reduced power to preserve the fuel supply. They came to a new heading that would carry them to Byron Bay and viewed their "how-goes-it" curve and hoped to fly the remaining distance to the safe haven. A "how-goes-it" curve is merely a check on distance flown, fuel burned, fuel remaining and distance to destination. It is a monitoring of the health of the flight so far, and a forecast of its future health.

They began the slow and gentle climb for altitude. If there was to be any chance of fuel starvation, they would be much better off up high; altitude would trade for time, and that time would give a better chance of finding a suitable landing spot in the tundra. Of course, this was only a precaution in case of fuel exhaustion, but they seemed to be doing well ...

As the hours slowly ticked away, the pair became a bit worried. A how-goes-it curve showed they might end up on the short end of the stick, but there was nothing to be done except press on. Finally, at 22:50 hours they could see the lights at Byron Bay. It looked like they would make it after all.

Suddenly their port engine coughed, followed shortly thereafter by the starboard engine. Their DC-3 was out of fuel and had become a thirteen-and-a-half-ton glider! Lundberg could see they were not going to stretch their glide to the strip at Byron Bay, but they would glide fairly close. This was preferable to being down in the tundra somewhere, miles from any help. They continued gliding toward the DEW-line site as they searched for a flat piece of tundra that might offer a chance for survival.

Lundberg spotted a fairly level surface, and he switched on his landing lights. He saw that the ground was flat and covered with snow. They had no way of knowing what might be concealed beneath that white blanket; he elected to land with wheels up. Bob considered this landing configuration much safer in case there were holes or ridges that might cause him to "stub his toes" and flip CF-CAR on her back.

The landing was surprisingly soft. The propellers, of course, did receive some damage. The belly was wrinkled and one propeller damaged the starboard tailplane, but no injuries were sustained. This was remarkable, considering they had 6,000 pounds of drill pipe in the cabin that could have killed them by sliding through the bulkhead onto the flight deck if a rapid deceleration had occurred.

In due course, personnel from the Byron Bay site arrived at their location with a Nodwel snow vehicle. But by then Bob and Gord had walked in to the radar site. Yet another Arctic adventure had come to a fairly happy ending.

* * *

Somebody told me that Cambridge Bay was sometimes fogged in on the open water of the bay. It was suggested that I practice during good visibility, making instrument approaches to the landing area. That 600-foot tower was located near the east shore of the bay. Flying northbound at eighty miles per hour, several hundred yards east of the tower and level with its top, I was advised to fly three minutes north, execute a rate-one turn to the left for one minute (a rate-one turn is one in which the nose of the aircraft comes around the horizon at three degrees per second). This would place me on final approach from the north, to the landing area in the bay.

At this point, I would descend at 200 feet a minute all the way to my landing. Every time I practiced this approach, it proved my timing was correct and I felt a certain confidence in using it, in case I ever needed to do so, under a 200-foot thick fog blanket.

One day as I was returning from Pelly Bay, I was advised by the air radio station that fog had rolled into the bay. At nearly the same time, one of our aircraft called as he approached from Yellowknife.

The other pilot told Aeradio that he would do a "fog do approach." I knew he would have to do a "for real" approach, the kind I had been practicing in good visibility.

The other aircraft arrived about a minute ahead of me and asked that I follow, giving him enough time to vacate the landing area before I landed. I followed about a mile behind him and watched as he executed a left turn. In short order, I, too, executed my own rate-one turn to the left and remained about a mile behind him. I watched as his aircraft descended toward the upper surface of that fog bank. Suddenly he simply disappeared from view—one second he was clearly outlined against the fog, and then he was simply gone.

I continued on my approach, descending 200 feet per minute, and I suddenly found the entire world had disappeared. The pilot of the other aircraft called. "Okay ... I've turned right ... I'm out of your way." I continued my descent and discovered that making this approach under ideal visibility conditions caused much less adrenaline flow than the present approach under zero visibility.

After what seemed a much longer period than 1 experienced under good visibility, my Edo 4580 floats gently brushed the surface, and they did so without a jolt; I became aware of white water spraying out to my left from my port float. Those practice runs had proved their worth.

When I had turned to the right, I had to reach a point no more than 100 feet from shore before it became visible. I felt my way along the shoreline; finally, our dock system came looming out of the fog.

Following this experience, I had to return to Yellowknife for a job of flying a young mining engineer to various points around the tundra. This flying job took place in 1967, during Canada's 100th birthday. My young mining engineer was Pat Parker. He had chartered my Beaver CF-NOT for a flight to several lakes. Although in his early twenties, two years' growth of beard gave him the look of an Arctic musk-ox. I was to fly him to each of the lakes, wait until he had surveyed mineral prospects and then fly him on to the next.

The first part of the day went rather well. Due to the tip-tanks with which NOT was equipped, we had no worries about running short of fuel, even with the deviations from a direct route to our ultimate goal of Cambridge Bay.

While Parker walked along ridges and valleys leading into the hills, I would stretch out beneath a wing and get a few minutes' sleep; once or twice a rain shower would force me to awaken and perhaps turn the aircraft around and secure a newly created shelter as I lay on the port float beneath the wing.

Parker wanted to go to a camp fifty miles east of Coppermine and eight miles south of Coronation Gulf. From our present lake, just east of Contwoyto Lake, we set course northwest in search of that camp. As we flew into the area west of Bathurst Inlet near our crossing of the Hood River, we encountered lower ceilings, and visibility began to deteriorate. After a while, I found a river that was perhaps five miles west of the camp we sought and turned east, hoping visibility would hold long enough for us to slip into that lake before the entire system clamped down tight.

In just a few moments, I found myself between layers that appeared to merge ahead. In a few hundred yards, we were in cloud and I was forced to change my direction 180 degrees. We were suddenly out of the cloud and were fortunate enough to locate the Wentzel River, which I had used as a check point. We landed only seconds before the entire system settled around us and those rocky hills 300 feet above us became lost in mist.

As the hours wore on, Parker constantly asked when we would be on our way. At first, my reply was polite. I pointed out the window and asked, "How can we fly out of this river when I can't even see the rocks on either side?" Parker insisted. He told me that Bob Shepherd had flown him in conditions that were worse than these. I knew Bob Shepherd and I also knew Pat's story was pure hogwash. But I replied, "Shepherd's a better pilot than I am!" After perhaps the twenty-fifth time that Parker asked me the same question about leaving, I became somewhat abrupt and said: "Pat, you may sit in that seat and discuss any subject under the sun. However, the very next time you ask when we can take off out of here, I intend on punching you right in the nose. We will leave this river when and only when I feel it is safe to do so. Until then, we will continue to sit in this very spot!"

After we consumed a few more raisins from my pack, weather conditions began to slowly improve. Within fifty-five minutes, we were airborne once again. We traveled north along the river, jagged rocky cliffs towering over us. I was looking for a "saddle" over the rocks, which would lead us to another river, a mile or two to the east. This saddle appeared on our map as a low passage among the craggy hills, and all at once, there it was. A much lower east-west path that would enable us to find that other river, and it, in turn, should deliver us to Parker's camp.

In short order we found the waterway we sought; I turned right and began the five-odd-mile flight south to the camp on the lake.

The river gorge narrowed, and before we had traveled half the distance toward the lake, the river was little more than a trickle and the valley was becoming more narrow. Because the cloud deck was lowering, I had no choices: 1 couldn't climb because I would not be able to stay within the narrow confines of this valley unless I could do so with visual reference to the terrain, and cloud made visual reference impossible.

In the end I was able to remain within the center of the valley, with no more than twenty feet to spare on each wingtip. We emerged from the river at about fifty feet above the lake, and the camp was just slightly to our left. The clouds were right on the water at the south edge of the lake.

We were having coffee in the kitchen tent when the crew told me they had heard us directly above them earlier but we had not been visible to them; when they heard the engine sounds abruptly stop, it was assumed that we had crashed. They were surprised, therefore, when we suddenly emerged from that narrow opening in the hills to their north.

Next morning, Pat was satisfied that all was as it should be in the camp and was anxious to carry on to our next stop. This next camp was on a lake near the south shore of Queen Maud Gulf, about eighty miles due south of Cambridge Bay. Upon our arrival I was inclined to just keep flying. The lake was shaped like an hourglass, although the eastern half was a bit larger than the western half. The entire lake was covered with ice, with only a narrow strip of open water along its northern shore, perhaps 100 feet wide.

I told Pat Parker that I was not happy taking my Beaver into that gap of water in case a wind shift might lock us in, perhaps even wreck our only means of escape from the place. Parker pleaded with me to give him five minutes with his crew and then we would leave. In the end, I acquiesced, but only to the point that would allow no more than five minutes. He told me he only needed that amount of time and went on to say he had to catch Pacific Western Air Lines DC-6 flight to Yellowknife and then on to Edmonton. Since this was Thursday, he'd be stuck in Cambridge Bay until Tuesday if he missed this flight.

The argument sounded logical, so I set up my landing parallel to the northern shoreline and pulled up beside a large flat rock on the north shore, which served as a docking system for that camp.

As I looked through the window, a very familiar figure came into view: Archie Talbot, and trailing behind him, Ingy Bjorensen. I popped open the window and shouted through the last few clicks of my magneto impulse starter. "Can you fellows tell me which way I should go to reach No Rum Lake?" I asked.

Archie looked at me as if seeing a ghost. It had been ten years since we had met, and he was now 2,000 miles from civilization. "Why, you old (expletive deleted)," he said, "Where did you come from?"

In keeping with their nature, Archie and Ingy had driven stakes into a pair of crevices to create a secure tie-down for old Beaver CF-NOT.

After five minutes passed, I told Pat it was time to go.

"Just another five minutes!" he pleaded. When that time had passed, I said we really must get going. This time, Pat said if I'd just have another cup of coffee, he would be right with me. Halfway through that cup of coffee, we heard a sickening shriek of tortured metal being dragged across a rocky surface. The wind had shifted—but none of us had noticed it from the relative calm of the camp atop the rocky hill. We rushed down the path to the big flat rock seventy feet below the camp. CF-NOT was being forced to her left, up the slope of rock. Her right wingtip was resting on the ice and her float bottoms protested loudly. The strong wind had slowly forced the ice sheet against the north shore. Fortunately, our Beaver was docked in a slight alcove; the ice was prevented from moving further as it was held up on the "horns" of the alcove.

We quickly checked the float bottoms—no damage. I ventured out onto the ice surface to examine the right wingtip and heaved a sigh of relief as I discovered about a sixteenth of an inch clearance between the wing and the surface of the ice. From atop the right float, using his crowbar, Archie and I finally were able to chisel enough candled ice out from the edge of the sheet so we could ease CF-NOT back to her original position at the edge of the flat rock. Her forty-eight feet of wing was level once again.

"Well," I told Pat Parker, "that ends our flying for today." But he would hear none of it. Pat had to be on that DC-6 from Cambridge Bay—today. "Pat," I told him, "the difficult we do right away; the impossible takes a bit longer." We were stuck until nature moved the ice out from around our Beaver. At this point, Pat borrowed the crowbar. It was his intent to go out on the surface and begin chipping the edge in order to open a channel from which we might take off. Archie, Ingy and I tried to dissuade him from this impossible task, pointing out the danger of such an attempt. But he insisted.

Archie told me, "Pat is young and his education is lacking, but I think that is about to change." Sure enough, within just minutes Pat had fallen through the ice and was clinging to the edge while his body rapidly lost its heat. The crowbar went to the bottom of the freezing lake, and Parker wanted help. None of us were willing to go to Pat's aid unless it became absolutely necessary. In the end, Parker was able to extricate himself from a situation of his own making. Some time after Pat had removed his wet clothes and changed to dry apparel, we heard the sound of that PWA DC-6 flying directly overhead, en route to Yellowknife and ultimately to Edmonton.

I suppose his disappointment at having missed that DC-6 flight to Yellowknife and Edmonton explained Parker becoming sullen with the other members of the camp, especially me, because we were totally unsympathetic. It was he, I told him, who insisted we make this stop. It was he, I pointed out, who had delayed our departure several times, even after I told him we'd have to go before the wind could move the ice and trap us. Later in the afternoon, the wind came about, and the ice sheet moved back to its original position, where we had a departure strip about 100 feet wide from which I was able to take off.

Throughout our eighty-odd-mile flight to Cambridge Bay, Parker sat in stony silence, refusing to even look in my direction. When we arrived at Cambridge Bay, a very strong east wind had sprung up before us; whitecaps covered the bay. Due to a very high wind, our landing run was quite short. It was impossible, however, to turn NOT. This resulted in being forced to "sail" her tail first toward the dock. With our Pratt & Whitney R-985 engine at idle and water rudders raised, I was eventually able to come alongside the dock without incident. Evidently, this maneuver made an impression on Pat Parker.

As he stepped from the left float onto the dock, he turned to me and said, "Well I'll say one thing, you sure know how to handle an airplane, anyway." Standing there on the dock at Cambridge Bay, I turned to Parker and said, "Pat, if you ever need to fly anywhere in future, please go to another carrier. If you insist on flying with this company, please ask for another pilot. You are a hazard to aviation. One of these days you'll pull that same stunt, trying to get some young, inexperienced pilot to fly in conditions beyond his ability by saying, as you said to me, that some other pilot took you through 'worse conditions than these.' I can visualize some young pilot taking you at your word, and you'll both end up dead."

I never knew just how much of that speech Parker actually heard, because the wind was howling around our ears. In the end, Parker walked away and our paths have never again crossed.

Two months after my encounter with him, Pat Parker chartered an aircraft flown by Cedric Mah, from Gateway Aviation. Cedric Mah had flown "the Hump" in Burma during WWII; he was a very experienced pilot. Weather was poor to terrible as they left Coppermine heading for Port Radium (Echo Bay) on the east shore of Great Bear Lake. Cedric Mah told me he knew they really shouldn't be in the air that day, but Pat Parker had told him, "Al Williams flew me through much worse weather than this, a couple of months ago."

Cedric said that comment influenced his own judgment. If Al Williams could do it, Cedric Mah could do it. Parker's claim was untrue, of course. I had decidedly not flown him through worse weather than they had that day. In fact, when we had found ourselves in lowering ceilings and poor visibility, I put down and waited for an improvement. By all accounts, their conditions were far worse than I had encountered when flying Parker around those Arctic regions.

Mah decided to go, and after a time their visibility became so restricted, they could only see straight down and very little in that direction. Cedric decided the only hope they had was to fly nose high, with 40 degrees of flap on their Cessna 185, with enough power to allow the Cessna to slowly sink toward the surface and hope for a lack of obstacles. After doing this for a while, he felt the floats contact water—rough water, with large waves, but he was happy to be down in one piece. They had come to earth on one of the Dismal Lakes, northeast of Great Bear Lake.

Just after the aircraft was firmly down on the surface, a particularly large wave roared up and their aircraft reared up on the wave nose high and began sliding back down from the crest. "We're sinking," came Pat Parker's panic-stricken voice. Cedric Mah told me, "When someone tells me we're sinking, I don't sit around arguing with him, so I opened the throttle to takeoff power." Because visibility was so poor, Mah had no way of judging that they were almost on shore. With full power the aircraft piled up on the ice and rocks on that shore. The floats were severely damaged and the aircraft reared up and began sinking, tail first, into the lake. The two men had barely enough time to grab their bedrolls and engine cover, together with their Emergency Locator Transmitter (ELT), as the waterline came within inches of the instrument panel.

For the following week or ten days, they were stranded on the shore of one of the Dismal Lakes. Cedric admitted one error—he had expected to hear any search aircraft looking for his downed Cessna. He misjudged; search aircraft were very high, seeking an ELT signal that was not being transmitted because Mah was "saving the battery" until he could hear an aircraft. In the end, a Pacific Western Air Lines DC-6 en route to Yellowknife made a swing toward the Dismal Lakes. When Mah heard that DC-6, he switched his ELT on. PWA spotted the signal at once and made a sweep directly over the downed men.

In short order, Willy Lazerisch of Northward Aviation was on his way to effect a rescue. The day of the rescue, I was leaving Norman Wells for the season as freeze-up was fast approaching. Somewhere between Wrigley and Fort Simpson, I heard Willy come on the frequency (5686 kHz) and say he was landing at the site. A few minutes later, Willy used the radio again. "I'm off Dismal Lakes … I have two passengers. They don't require medical treatment but one of them says he needs a nurse."

It saddened me to learn a few days later that on that same day I had been approximately eighty miles east of Little Dal Lake, Bob Shepherd had been killed just north of the lake. Two openings in the mountain range north of Little Dal look very similar. The first opening leads to a box canyon, while the second will take an airplane to a parallel valley about six miles east. Unfortunately, because low ceilings and poor visibility made it difficult for Bob to see the proper entrance, he had turned into the first one. He was unable to execute a turn in the narrow confines of that box canyon; his Beaver hit the eastern wall in a ninety-degree bank and ended up crashing into the north wall of the canyon.

Pat Parker, under different circumstances, might have been with Bob Shepherd on that final flight, from which no one survived.

I heard Cedric Mah's story about Parker when we both found ourselves weather-bound in Fort McMurray at season's end. It had previously occurred to me that perhaps I had judged Parker rather harshly. I changed my mind when Cedric Mah had finished his tale; I concluded that my judgment had been much too mild.

The following day, after depositing Pat Parker on the Cambridge Bay seaplane dock, Bob Warnock, who was one of Northward's pilots at the base there, asked me to fly Otter CF-CZP on a route check to Pelly Bay, just over 400 miles to the east.

Bob Warnock had been flying practically day and night—we were in the land of the midnight sun—and he was tired; Bob hoped that if I could take over some of that flying, he would be able to find some rest.

Bob's normal load was 2,700 pounds in the cabin. The gasoline tanks were filled to capacity (178 gallons). Two forty-five-gallon drums of 80/87 aviation gasoline were placed in the cabin behind the left-hand seat, just aft the flight deck. At 7.2 pounds per gallon, we carried 1,929 pounds of fuel. Since Pelly Bay had no fuel available, it was necessary that we carry sufficient fuel for our return flight as well.

I never stopped to figure it out precisely (which I should have done), but we were several hundred pounds over gross weight, which I knew was 7,967 pounds. With our present load the tip of the rudder, which should have been sixteen feet above the waterline, proved less than sixteen feet. There was ice at the far side of the bay, and we had to lift off before reaching the edge of that ice sheet. By applying the lessons I had learned from Shorty Holden away back when I flew with Parsons Airways Northern, I was able to become airborne before reaching the ice, although Bob Wamock had a few moments of uncertainty. We surely needed all fifty-eight feet, ten inches of wing during that takeoff, and all of our 600 horsepower as well.

Departing Cambridge Bay for Pelly Bay, we climbed at only fifty feet per minute and passed just north of the DEW radar site on Jenny Lind Island. About forty miles west of the Gladman Point radar site, on King William Island, I could see a cloud deck at nearly our altitude of 5,000 feet. I decided, rather than relinquish any of that hard-earned altitude, I would climb on top of that thin deck of clouds, because I could see sunlight, reaching the ground, on King William Island beyond the cloud layer.

As we climbed to near the top layer, Bob Warnock awakened from his sleep. He glanced left and right like some trapped animal and then he pointed toward the ground. "Take her down!" Bob shouted.

"No," I said, "We are nearly on top of this local cloud. Don't worry, we're right on course and there is nothing to worry about." Just a few seconds later we broke out into bright sunshine, and Bob relaxed as we spotted King William Island beyond the eastern rim of the cloud.

Having crossed King William Island, we over flew Gjoa Haven, on the southeastern coast at 68 degrees 38 minutes north and 95 degrees

57 minutes west. Roald Amundson named Gjoa Haven after his ship *The Gjoa* (Joe) after he arrived there in 1903. Amundson wintered there two years. Flying over the area, one wonders about the kind of protection available to them, because the land scarcely rises beyond forty to fifty feet above sea level; winds can sweep in from Erebus Bay, nearly a hundred miles to the northwest. The highest point on the island is a flat-topped hill, just 450 feet above sea level, which is about twenty miles northeast of Gjoa Haven.

We continued on to Pelly Bay, where we off-loaded our cargo and refueled CZP from the two forty-five-gallon drums we carried. We were soon winging our way back to Cambridge Bay for a second trip.

Our return trip to Cambridge Bay was a bit easier because of a tremendous improvement in performance as Otter CF-CZP climbed to our cruise altitude without a struggle. Since Bob was so tired, and because I had proved I could find my way to Pelly Bay and return, Wamock stayed in Cambridge Bay for a second trip, which I made by myself. Without Bob Warnock's body on board, CZP climbed a bit faster and the entire trip went off without a hitch. Now I had flown about 1,600 miles (400 miles each way on two trips). I was becoming tired, especially with the great noise generated by that Pratt & Whitney R-1340 engine pounding in my ears for over fifteen hours.

Upon my arrival back at Cambridge Bay, a message was waiting for us to bring CF-CZP to Yellowknife at once. Well, Yellowknife was some six hours away. Bob had still not completely caught up with his lack of sleep, so he elected to have me fly CF-CZP while he got out his bedroll and curled up on the cabin floor for more sleep.

We had flown well past Contwoyto Lake to within about 100 miles of Yellowknife when we encountered heavy turbulence. Otter CF-CZP wallowed around like a sick cow trapped in quicksand; its wings would dip to left or right as much as 80 degrees while I valiantly fought to maintain an even keel. Warnock awoke almost immediately and came staggering, as best he could, to the flight deck and took the right-hand passenger seat. Because of my concern for possible structural failure, I reduced power to twenty-six inches of manifold pressure and 1,650 rpm for just 300 horsepower; of course old CZP had slowed to about eighty-five miles per hour. Bob Warnock began to feel the effects of the battering that we were experiencing; he reached for an airsick bag while he swivelled in his seat and put his head down, preparing to be sick. He also reached up and increased our power setting, because he was anxious to get out of this turbulence as quickly as possible. I could not take the chance on structural failure from the heavy battering we were taking, so I reduced our power setting immediately.

I probably would have experienced the same near-airsickness he was experiencing (Bob never did actually throw up) had I been in his position, but I was much too busy to become airsick. By and by, when we were about thirty miles out of Yellowknife, the atmospheric violence ended and we were able to resume twenty-nine inches of manifold pressure and 2,000 rpm and continue our flight using a full 400-horsepower cruise setting, around 67 percent of takeoff power.

Upon our arrival at Yellowknife, we discovered there had been a misunderstanding. Otter CF-CZP wasn't required. I never found out how the information had become so garbled, but Bob Warnock just refueled and returned to Cambridge Bay. I never returned to Bob's base, but I had flown Otter CF-CZP nearly twenty-two hours over the past twenty-four hours. I barely dragged myself upstairs to the crew room where I slept around the clock. The Otter produces so much noise that the din itself is very tiring.

A few months later, Bob Warnock learned that he had leukemia; it was not long after he had taken a corporate flying position. (He would die within a year of taking that new job.) When Louise and I met him a short time after he began flying as a corporate pilot, he bemoaned the fact that the company had insisted on his shaving his handsome beard, which had suited him so well. Bob always wore a pair of Wellingtons and a blue roll-neck sweater, which matched the brilliant blue of his eyes.

On July 21, Bob Shepherd invited me to go for a test ride in another Otter CF-IKK. Bob was a friendly man, much like Oscar Erickson had been. They shared something else in common. I noticed that during our takeoff Bob would nod his head and the closer we approached liftoff speed, his nodding would become faster. His knuckles were white (just like Erickson's had been); his heels came off the floor as he put heavy pressure on the rudder pedals. Once CF-IKK was in the air, the color returned to Bob's knuckles and his heels rested gently on the floor and there was no longer a bobbing of his head. Poor Bob Shepherd was obviously nervous while flying. I expect that may have been a contributing factor to his death.

21 The Norman Wells Experience

The day following my ride with Bob Shepherd, I was dispatched to Norman Wells on Pacific Western Air Lines' DC-6 service. The flight from Yellowknife to Norman Wells was flown above the overcast and so I could not see the terrain until we descended to land on the air strip at Norman Wells.

Upon our arrival, I was installed in the company house where senior pilot John Davids was staying. At that time we had a very well-kept building, for which I believe full credit must go to our base manager Lee Work (pronounced like Cork). Periodically, Lee would stop by. I feel quite sure that those short visits were done more to keep an eye on how we were looking after the place than out of friendliness. Our building was next door to the local liquor store. As a nondrinker, I never paid the place a visit.

Lee Work had hired an elderly Inuit man named Charlie. He looked after any little repair jobs for which we might have a need, and he tended the few flowering shrubs which adorned the front yard, as well as touching up weathered paint. An Inuit lady was also retained to come in once a week to do the cleaning. We did our best to keep the place reasonably clean, but because we spent most of our time behind the controls of a de Havilland Beaver or Otter, housework became less of a priority. We were well equipped with a deep freezer, plus washer and dryer. Because of the use of "utilidores" for hot water central heating, which came from the Imperial Oil refinery, we had only hot water for our building, including the toilet. (The utilidore is a well-insulated structure stretching all over town. Hot water for heating and cold water for drinking and washing run side by side inside these structures, because of permafrost underground; the heat would soon melt the permafrost, a disastrous situation.)

It saddened me to read many years later, in Robert S. Grant's book *Bush Flying: The Romance of the North*, that when Lee Work was no longer in charge of the Norman Wells Northward Aviation base, the house was trashed. While I was there, during the summer of 1967, we felt a certain pride in that home. After John Davids left, our new engineer, Malachi McCarten, moved in and we continued to keep the place up.

John Davids was one of the finest gentlemen with whom I had ever flown. He was a heavyset man, and I shall always remember him as being kind when someone had stolen my microphone and earphones in Hay River. Most pilots prefer their own personal pieces of this aircraft gear, but I was left without either item. John Davids was under no obligation to do so, but he lent me his personal "mike" and "phones." John said he could get along without them because he was not a regular line pilot; he would fill in from time to time when it became necessary, but he didn't fly as a matter of routine. I guess he would have borrowed a set if needed for one or two flights.

A Canadian registered de Havilland Otter CF-VVY had been flying in Africa (the same airplane which Bob Lundberg had used in giving me my "wheel endorsement" ride). Some brave soul took on the job of flying her out of Africa, across the Atlantic Ocean to South America. Extra fuel tanks had been fitted into the cabin for that flight, and the aircraft was operated on wheels, although I suppose a landing at sea would have been a disaster without regard to what kind of "feet" she wore, wheels or floats.

Eventually CF-VVY was delivered to Miami, Florida, and it became necessary for someone to pick her up and ferry her to Edmonton. I was briefly considered for the job, but in the end John Davids was chosen for that ferry flight. As fill-in pilot, he had drawn the job while I was operating out of Yellowknife and Cambridge Bay.

Norman Wells stands on the east bank of the Mackenzie River, and DOT Lake (named after the location of Department of Transport radio installations) is about one mile south of the settlement and airport. Because of the overcast condition and prevailing wind on the day of my arrival, our DC-6 flight approached to land from the north, with the result that my first sight of the place produced a 180-degree shift in my navigational orientation — a perception I was never able to correct during the time I spent there. I wrongly assumed that we were landing from the south and that Norman Wells was on the west bank of the river and that DOT Lake was a mile or so to the north.

My disorientation had no effect on my ability to leave Norman Wells and return at the end of each flight, but it was disconcerting; I found that my compass orientation was restored once I flew beyond the Norman Range. Each time I left the immediate vicinity of Norman Wells, I was able to reorient myself in the valley.

My first flight from Norman Wells, shortly after my arrival, was to Fort Good Hope, with Joe Herman of the CN Telecommunications Division; Joe was also engaged in the Mackenzie Valley upgrade of the CN telephone system. The ceiling was very low, and I found that I had no choice but to fly within the confines of the river. It was an interesting introduction to flying out of that locale.

Next day I had to fly to a tourist lodge under construction on Ford Bay, just off the south side of Smith Arm on Great Bear Lake. Nothing very exciting happened during that trip, but it was notable that I was unable to taxi closer than 300 yards from shore because so much ice was floating in the bay. The men had launched a boat and paddled around and between closely packed ice floes to reach my Beaver.

The next morning, I had a flight to Colville Lake (located within the Arctic Circle). Father Brown at Colville Lake was somewhat of a character in the area. Among other things, he was an accomplished dog sled master, an oil painting artist and an accomplished builder with logs. His paintings adorned the interior of his log church. Of course, the Hareskin (Sahtú Dene) people who lived there carried out most of the actual log construction of the church and cabins for rent to Arctic tourists. Brown operated a sort of tourist lodge, and although he didn't own an airplane, I was told he was a pilot as well.

Father Brown had an eye for the ladies. I made several trips with female school teachers and nurses, who would often "spend a night or two" with him. The priest was a sharp entrepreneur as well; he sold me a Brazilian *um cruzeiro* note (valued at one-tenth of a cent) for twenty-five cents. I say that he sold it to me because I had purchased some tobacco from his little store and spotted the Brazilian note in his cash drawer. I commented on it, and Father Brown told me that his brother, a priest in Brazil, provided him with these notes.

Father Brown said, "Here, you can have it." I picked up the *um cruzeiro* and he added, "That's only twenty-five cents." At that point I had little alternative to buying it. Live and learn. In future, I made no comment about anything I saw in his store, unless I had a purchase in mind. I still have that *um cruzeiro* note.

There were many routine flights, often serving Joe Herman at Fort Good Hope or flying other CN Telecom crews into and out of various telephone repeater sites along the Mackenzie Valley. There are too many names of northern pilots to recount in this book. I have mentioned Bob Wamock, with whom I flew out of Cambridge Bay. I should give honorable mention to Paul Hagedorn as well, who flew with Northward Aviation during my stint with that company.

It was Paul Hagedorn who opened up the base at Cambridge Bay in the 1960s and who, when he left the north, started his own flying service under the name Geographic Air Surveys, flying photographic missions out of Edmonton. Paul spent more than twenty-five years in the Arctic. I should also mention the famous bush and Arctic pilot Ernie Boffa, who was flying a Norseman out of Sawmill Bay during my tour of operations in the western Arctic, as well as Mike Zupko, who was operating out of Aklavik at the time.

Recently I spoke with Gerry Bolding, who recalled a story I had shared with him and his supervisor, R.C. Jones, when they were with the Medical Electronic section of the Royal Alexandra Hospital in Edmonton. The incident took place while I was based at Normal Wells.

After I had all my teeth removed and dentures fitted, a short time prior to becoming employed with Northward Aviation, my gums shrank and I was bothered by loose-fitting dentures. One day, I found a tube of silicon rubber sealing compound in our maintenance shop; I smeared a thick layer of it in my denture plates. I noted a very strong vinegar taste as I fitted the plates in place, but in a few seconds, they were firmly set. For the remaining flying season, my dentures fit very well.

When I left Northward Aviation, I told a man named Wally Phinney about my solution to loose dentures. Wally believed this might be a solution for his own loose dentures but was disappointed. The chemical which gives a vinegar taste to the compound attacked his gums. He was unable to wear his plates for several days. I considered my denture solution just another "bush pilot fix"—one that had worked out very well for me.

22 The Yukon

On July 26, 1967, I began to fly for a new client, Petro-Par of Canada Limited. Petro-Par was a French company engaged in mineral exploration in the Yukon. The two French principals were Bernard Pluchett—who was fluent in English and Francois Dufaud—who spoke only one phrase in English. At breakfast, lunch and dinner, Francois would invariably be a few minutes late; he'd arrive after everyone was seated and, together with his grand entrance, he would announce in a heavy accent, "Good evening ladeez and gentlemen." There were no ladies present, and I'm sure Francois said it just for effect. He always got a laugh and therefore was encouraged to repeat it at the next meal.

Warren McKenzie was a mining engineer who came from Calgary, and a very fine fellow he was, albeit a bit quiet. Jack Leroux was our camp cook, who doubled as our helicopter engineer; Jack was out of Africa and held a helicopter-only aircraft engineer's ticket. Only a handful of such licenses had been granted at that time. I imagine that has changed during the past twenty-nine years.

Our helicopter pilot was Tomatsu Komiho, who had only recently arrived from Japan. I remember his first breakfast at the camp. Jack Leroux had cooked up a large pot of Quaker Oats—a dish that had never found its way to Komiho's table setting in his native land. Poor Komiho didn't know what to make of it; bear in mind that a lump of porridge, with its gray and rough texture staring up at the uninitiated for the first time, must be intimidating. Poor Komiho found the porridge as unappealing as anything he had ever seen. He had no idea how to eat it but watched and then mimicked the others.

To his surprise, when he took his first spoonful of oatmeal with milk and brown sugar, he loved it. Future breakfasts that featured oatmeal were a treat to Komiho, although he also enjoyed bacon and eggs just about as well.

It was Warren McKenzie, I think, who lent Komiho his fishing rod, something else new to Tomatsu. Japan is famous for its fish—but fish are netted, not caught with a rod and reel. The day Tomatsu Komiho got his first rainbow trout high in the mountains, in a very cold mountain stream, he was beside himself. "Big thrill," he said over and over. "Big thrill!" From that moment on, Komiho was a devoted angler. Every day, he would bring home rainbow trout for our evening meal.

We established our main camp on a lake, which for obvious reasons we called Necessity Lake. It was a necessity to operate from that lake because it provided us with potable water and was the only lake large enough for our red, white and blue Beaver CF-GQN. At least there were no other such lakes in close proximity.

The helicopter would fly the prospectors from a forward camp eighty miles west of our base to service their needs during the day. CF-GQN, on the other hand, was used to transport everyone, except Leroux, to that forward camp. I would fly them out in the morning and check their fuel cache; if needed, I would return with a load of ten-gallon kegs of 80/87 aviation gasoline for the helicopter, returning to pick them up late in the afternoon and fly them back to our main camp for one of Jack Leroux's exquisite meals. Periodically, I would fly Jack out to the forward camp so he could do routine maintenance on the helicopter. It was too expensive to fly the helicopter the 160 miles each day out to and back from that forward camp, when Beaver GQN could handle the job at far lower cost.

One day we found ourselves running short of ten-gallon kegs. Jack Leroux and I decided we should take GQN back about halfway toward Norman Wells to "Cache Lake," whose waters were the color of the bluish gray gravel that cascaded down the north slopes of the mountain range. Actually, we only called it Cache Lake because of the fuel cache we had established there; Cache Lake was quite small and its waters shallow. Its bottom was known by locals as "loon shit" because of its thick, spongy consistency.

Our gasoline cache was located on the south shore, just off a small bay at the southeast corner of the lake. Because the wind was from the north, I taxied toward the south shore about 100 feet to the west of the little bay. My plan was to turn left into the bay and shut down my engine; I'd then raise my water rudders and allow GQN to sail back into the bay until we were resting on the bottom. However, I ran into a problem. The water was dark and cloudy and I was unable to see the bottom; therefore

I didn't see the shallow stretch in our path until we came to a sudden but gentle stop, as the keels of our floats became stuck in the mud.

I shut down my engine, and Jack and I removed our shoes and socks and rolled up our pant legs as we stepped into the mire. I could feel the soft, rubbery bottom oozing up between my toes; it was decided that we could probably lever our way out of the muck if we could put a couple of pries under the forward edge of the front spreader bar. I removed the ax from its storage place, and Jack went into the woods not far from the cache of gasoline drums; Jack felled two jack pines about ten feet tall and removed the branches and twigs; these were placed under the front spreader bar, and we exerted a mighty heave. GQN moved about three feet, but when we removed our pries to try for a new purchase, the mud just sucked our floats right back to their original position.

Any help was at least 200 miles away, but we knew that no one else could do anything that we were not capable of doing for ourselves.

At the time, I was a smoker. I sat on the left float, smoking, while I pondered our situation. Jack Leroux became quite angry, since he had his evening meal to plan and prepare. I was also anxious to get going, since it was approaching my pick-up time for the crew at the forward camp.

"Come on," Jack yelled, "help me pry with this pole; we've got to get going."

"Jack," I said, "we've tried that approach and it just doesn't work. We've got to figure a better way to get this airplane turned around." Jack muttered a few uncomplimentary words and returned to his levering and sweating and swearing. "Jack," I asked him, "how does a caterpillar tractor or an army tank turn?"

He stopped prying with his lever and turned to me. "They stop one track while the other one is going, but what does that have to do with anything? This airplane is on floats, not tracks."

"Right on, Jack," I said. "But if we could stop one float from going forward and allow the other to move, we would turn her right around, correct?"

Jack didn't get it at first. "One of us will have to get onto the left wingtip. That will force the left float to bind even harder. At the same time, the right float will lift part way out of the mud. We have a Pratt & Whitney R-985 up front which can deliver 450 horsepower. We start the engine and open up to near takeoff power; she should come right around." Jack was not convinced, but he had to admit that prying on those floats wasn't working.

"I'm not going on that wingtip," he said. "I'm a private pilot and helicopter engineer, not ballast."

I told him, "Okay, Jack, I'll go to the wingtip and you open up the throttle to about half takeoff power. Watch me, though—I'll move my hand in a circular motion for more power. I'll drop my hand straight down when I want you to chop the power. Okay?"

I climbed up onto the left wing and moved right out to the tip. Jack, in turn, started the engine, and after an appropriate warmup, began to advance the throttle. Just a bit short of takeoff power, GQN began to swing to the left. I dropped my hand and Jack pulled the throttle lever back to idle. GQN continued swinging left to turn 180 degrees. When I resumed the left-hand seat, I used GQN's engine to pull us free. In short order we had loaded twelve kegs of 80/87 Avgas and were on our way back to our main camp.

After our departure from Cache Lake, we headed west back toward our base camp to arrive in plenty of time to perform our duties. Not long after takeoff, Jack asked me about a statement I had made during the previous day; I had said I could, and often did, take GQN off using only cruise power. Although Jack had a private pilot license, he had only flown Cessna 182 and smaller aircraft and was skeptical about my claim. I offered to demonstrate with the twelve kegs we had on board. This did not constitute a full gross load, but GQN was fairly heavy for such a demonstration, using only cruise power for takeoff.

We kept our eyes open for a suitable stretch of water where the demonstration could be carried out. We soon found the perfect place.

The lake ran east and west with about a five-mile-per-hour east wind. I dragged the lake, seeking submerged rocks and deadheads—hitting one can spoil a pilot's whole day. The landing was uneventful, and I showed Jack how to set up for a cruise power setting. At 1750 rpm and twenty-nine inches of manifold pressure, this would yield about 260 horsepower. I had Jack set our power at these figures so he could not claim I had cheated. As he began applying power, I pulled the control column back into my chest and watched the bow wave; when it stopped moving aft, I came rapidly forward, but only an inch forward of neutral. GQN came onto the step, and when she did I moved the control column slowly forward and back until I could feel a pressure increase between the seat's cushion and my back, just as I had learned from Shorty Holden. This told me when I had achieved the ideal float-to-water angle for best acceleration. We left the surface; Jack was amazed. Not bad for 57 percent of takeoff power.

I said that there was nothing magical about what I had done. I had only used an application of science to reduce expenditure of energy by allowing our floats to work more efficiently against the surface of the water.

"I can teach you to do the same, Jack." I told him about the lessons that Shorty Holden had given me on the care and feeding of seaplanes.

Leroux had never flown anything larger than a Cessna 182; he was reluctant to try his hand at a de Havilland Beaver, with its forty-eight feet of wing, its 450 horsepower Pratt & Whitney R-985 engine, weighing in at 5,100 pounds. Add to this the fact that Leroux had never tried flying a seaplane in his entire flying experience.

Nevertheless, we traded seats. I handled the power settings for him; I had to coach Jack through getting the control column right back into his chest during the early stages of takeoff, but then he mimicked my rapid forward thrust of the control column, and GQN came onto the step.

Again, I had to coach him to find the perfect float-to-water angle, but once he had found it, he managed to hang onto that elusive angle. He used nearly twice the amount of room I had used, but he did get GQN into the air with only cruise power settings.

His second takeoff attempt was better, as he needed less coaching. This time Jack used only about 20 percent more room than I had used.

His third and final attempt was better still; Jack managed to get airborne in nearly the same distance that I had used. Once we were both satisfied that a de Havilland Beaver can be made to fly using much less than its takeoff power, we turned on a heading that took us back to our base camp. Jack went into his kitchen tent to prepare our evening meal; I flew to our forward camp with the twelve kegs of 80/87 aviation gasoline and picked up the crew, leaving those kegs of fuel for the helicopter.

On July 30, the two company executives wanted to fly to Inuvik for a meeting and then on to the Carcajou Lake area to find a new camp site. We located an ideal spot for their future camp and then flew on to

Fort Good Hope, returning to our own camp later in the day. We had spent eight hours, twenty minutes in the air that day.

Later, when the camp had been moved to a clear, cold stream on Carcajou Lake, I had occasion to fly a supply of gasoline drums into their camp using de Havilland Otter CF-CZO. A very strong wind was blowing from the west, and whitecaps crested the tops of the waves as I landed my Otter. I sailed toward shore; but before reaching the shore, my floats ran aground on the sandy bottom and I found it necessary to push the drums out of the airplane so they could drift ashore. Gasoline weighs 7.2 pounds per gallon, while water weighs ten pounds per gallon. The drums floated closer to the water's edge than I could maneuver the floats of CF-CZO. While it worked very well, it seems a bit incongruous to toss full forty-five-gallon drums into a lake.

At some point during my service flying for Petro-Par, I had to return to Norman Wells for routine maintenance on Beaver GQN, and to fetch a few more kegs of gasoline for the helicopter. At about the halfway point I saw fog building on the hills and rolling down into the valleys. This was arid terrain, and I was looking for one of three small lakes near my position. Suddenly I spotted one of those lakes, but I knew there was only one that was long enough to accommodate a Beaver. I could not be certain which lake I was approaching, because fog covered more than half of its surface. Norman Wells reported a fog buildup was in progress, and looking back in the direction from which I had come I was sure fog was building there as well. My best chance was to get out of the air as quickly as possible.

I lined up with the visible part of the lake and committed myself to landing there, hoping that this was the lake that was long enough for GQN. As we touched down, very near the approach end, I hauled back on the control column, digging the float heels in so as to shorten the landing run. I was well into the fog by the time I nearly stopped; 1 kept expecting to run out of room, but fortune smiled on me that day; I had found the correct lake. I shut the engine off and allowed GQN to drift, tail first. Only a very light breeze was blowing from the east, and we slowly drifted toward the west end of the lake.

I had planned on going ashore, but when GQN had drifted within a few feet of shore, I stepped out onto the port float, took up the paddle and knelt down to test the bottom. The paddle touched the muddy bottom just a couple of feet beneath the surface, but when I pushed on the paddle, it kept sinking. Had I stepped off the float, I might very well have sunk deep enough to have drowned in that murky, muddy water.

For the next three days I was stuck on that muskeg lake, unable to go ashore and stretch my legs. I had about fourteen empty kegs in the cabin, and these were removed to the left float and secured in place with a length of rope; I needed that space for sleeping. I had plenty of canned goods (including several cans of peaches) as well as a loaf of dried bread. I always carried a loaf of bread with its bag open so the bread would dry quickly and not go moldy. I could easily soften the dry bread with pork and beans or other canned goods. I took my liquid from canned fruit I kept on board.

On the third day, the skies cleared, the wind blew the fog away and I found myself on my way back to Norman Wells and an overdue bath and a shave. A few days later, I flew Petro-Par to Arctic Red River and then to Inuvik. 1 can't recall the purpose of that trip. I mention it here only because I made an entry in my pilot's log that shows five hours, fifteen minutes of flying that day.

On August 3, using Beaver GQN again, after flying gasoline to Carcajou Lake in Otter CZO, I flew seven hours, going to Carcajou Lake, Cache Lake, June Lake and back home to Norman Wells. I was sorry when I had to leave the Petro-Par crew; we had developed a bond between us. I missed Francois and his recital before every meal, "Good evening ladeez and gentlemen."

Bush and Arctic Pilot

23 Mercy Flight

It had been a long day on August 16. My first flight began at 04:35 hours, and the pressure hadn't let up all day. My aircraft was de Havilland Beaver CF-GQN. At 19:47 hours, I had only to fly to the dredge at San Sault Rapids, just fifteen minutes by air along the Mackenzie River, north of Norman Wells. I had flown to Fort Norman and Fort Good Hope, to the Chick Lake repeater (a CN repeater building) then up to June Lake, in the Backbone Range. A trip to Kelly Lake and return was my next-to-last trip of the day. We had a few groceries to deliver, together with mail, to the dredge. When that trip was completed, I would be free to go to our company house, which I shared with engineer Mai McCarten. I'd have a bite of dinner before falling into my bunk for much-needed sleep. I had already been in the air seven hours and forty-five minutes during the day, so far.

Less than five minutes into my flight, I was called on HF radio by Air Radio. "GQN, Norman Wells ... company requests your return immediately."

I must have forgotten something, I thought to myself, as I began a turn back to base. "Roger ... GQN," I said.

Several people were waiting for me on the dock. I was told that we had an emergency flight to Inuvik. Mike, who was in his mid-seventies, had been working with the diamond drilling crew some forty-five miles away along the river. A core barrel had fallen on him and had severed a goodly portion of his right hand. When contacted by radio, Inuvik had advised the camp to bring Mike to Inuvik Hospital as quickly as possible; there might be a chance microsurgery could save the hand, if not completely restore full usefulness. The hand was placed in an envelope, and both Mike and his severed hand were transported to Norman Wells. The speedboat they used was very fast, but two hours had gone by before they arrived at our seaplane base.

If they had called us first, I could have shortened that time considerably; as it was, it would take an additional three hours and five minutes or more to transport Mike from Norman Wells to Inuvik by air. Because my last flight of the day was only thirty minutes, I had not bothered to refuel, as there was still more than thirty-five gallons. Now, with this Inuvik flight, I'd need full fuel tanks. I would have about four hours of fuel, more than enough to make it to Inuvik with—as John Davids called it— "grandmother gas" left over, in reserve.

Because of the nature of Mike's injury, I readied GQN for flight as quickly as I could. Emile Jackson, a member of their drilling crew, had brought Mike this far; he would, naturally, come with us as we made our way to Inuvik. In our hurry to get underway, I neglected to rid myself of the remains of several cups of tea which I had consumed at my last stop, prior to my flight to the dredge. It was to become an important factor as this flight progressed.

"Norman Wells, GQN ... out on the hour, off at zero five. We're climbing on course to flight level five zero, in the clear ... estimating the Grand View at twenty-five next hour ... Inuvik, GQN." This told Air Radio that we had departed DOT Lake at 20:05 hours, we would be cruising at 5,000 feet and should arrive at the Grand View, on the Mackenzie River, at 21.25 hours, with Inuvik being our ultimate destination. These types of abbreviated radio exchanges imparted a lot of information.

Just beyond Fort Good Hope, I needed to go to the bathroom. I reached for one of the airsick containers to urinate into. Try as I would, it was not possible for me to pass water sitting down. I would have to stand or be in a semi-standing position.

GQN had no autopilot, and Jackson had never been in a de Havilland Beaver in his entire life. He swore he knew nothing of flight control and would almost certainly kill us all if he made the attempt. After several demonstrations and a bit of practice, Jackson was able to maintain a fairly level flight while I struggled out of my left seat and positioned myself in a standing crouch configuration just behind Emile's seat. I can truthfully state that Mike was the most stoic individual I've ever met. He must surely have been in terrible pain, because only basic first aid had been available at the drill site. His hand was bandaged neatly, but he had received no pain killers beyond a few ASA tablets for more than two hours.

Periodically Emile would turn to him and ask, "How're you doing, Mike?"

Mike's reply was always the same. "I need a cigarette." The only help I could offer Mike was to get him to Inuvik as quickly as I could. Finally, around 23:15 hours, we landed and docked at the company air base. An ambulance was on the dock, and Mike was whisked off to Inuvik Hospital immediately while Emile and I arranged for a bunk in the Northward Aviation crew room.

I had a pick-up scheduled for 06:30 next morning at Kelly Lake, a few miles from Norman Wells, so I told Emile to be prepared for an early start. We refueled GQN and settled down for a short sleep. At 04:00 I was up, and before we left Inuvik, Emile had phoned the hospital; we both felt the disappointment when he was told too much time had elapsed between the accident and our arrival for old Mike's hand to be saved. We had to leave Mike in Inuvik Hospital to convalesce. I never heard anything more about what became of the poor old fellow.

Emile and I arrived at Kelly Lake and made our pick-up of passengers in time to put them onboard the DC-6; later, when we both were living back in Edmonton, I was fortunate enough to meet the man after whom Kelly Lake had been named. His name was Pat Kelly, and he worked in the mobile radio shop for the Alberta Government Telephones Company.

Pat had been the pilot-in-command, flying a survey crew out of Norman Wells. They were mapping and naming various lakes. Their job was nearly complete when Pat flew over a ridge; there before them lay a large and beautiful unnamed lake. Since they had used up all the names on the list from Ottawa, someone suggested they call it Kelly Lake, in honor of the pilot. Kelly Lake is nestled in the Norman Range near Norman Wells and is not easily seen from the air unless one's aircraft flies directly over the valley.

Bush and Arctic Pilot

A. R. (Al) Williams

24 Mackenzie Valley and Great Bear Lake

The Canadian National Telecommunications division was busy with the Mackenzie Valley upgrade (improving the communications system of telephone and broadcast facilities). Lee Work and I were sharing de Havilland Beaver CF-GQN and Otter CF-CZO, but I did nearly all of the flying for CN. It was just happenstance that each time the CN crew needed a Beaver, it was my day for CF-GQN; when they needed the Otter, more often than not, I happened to be flying CF-CZO (or CF-VVY).

Nearly thirty years later, Lee Work was asked if he remembered me and responded, "Oh yes, Al flew the Beaver for us one summer."

The truth is that he felt somewhat irked when I arrived, because chief pilot Bob Lundberg had decreed that we would share the Otter and Beaver equally. After nearly thirty years, Lee was not willing to admit that we had shared those two aircraft.

Sharing the planes stood to reason, from Bob Lundberg's point of view. I had many hundreds of hours on Noorduyn Norseman aircraft, which used essentially that same Pratt & Whitney R-1340 engine (although the Otter had a gearbox to reduce the propeller revolutions, not included in the Norseman installation). Nobody could get insurance coverage on Norseman flying until they had at least 1,000 hours on other fixed-wing aircraft. I never heard of any similar restriction for flying the Otter, and that is likely because that aircraft is so much more user-friendly than the Norseman.

One day, Larry Carter, Bob Bogart and Jack Keenan needed to fly to Wrigley, some 176 miles south, along the Mackenzie River. Jack Keenan always sat reading comic books whenever he rode with me; I thought it was because he found flying a total bore. I was to learn that this was not the case at all. Jack Keenan was terrified of flying—a fact that no one had told me.

Larry Carter was sitting in the right-hand passenger seat by my side, while Bob Bogart sat directly behind him and Jack Keenan was in the seat immediately behind me. Carter was curious about anything to do with aircraft; he'd ask the cubic inch displacement of our Pratt & Whitney engine. The number of the Pratt & Whitney used in the de Havilland Beaver was R-985, meaning that it was of radial design—it had nine cylinders arranged in a radial pattern, and it had a displacement of 985 cubic inches. The engine in Norseman and Otter aircraft was an R-1340. It too was a radial, with a displacement of 1,340 cubic inches.

Larry asked me what a stall was like in a de Havilland Beaver. I looked around at Bob Bogart, and he nodded his approval for me to demonstrate. I also looked around at Jack Keenan, but he was engrossed in a comic book, so I assumed a stall would not be a problem for him either. Had I perceived that Keenan would be terrified during such a maneuver, I would not have carried it out. I pulled the nose of GQN up, quite steeply, allowing her to reach her stall speed, and her right wing dropped quite steeply, recovering in just a second or two. I restored cruise power and we resumed level flight.

I looked around at Bob Bogart and he was all smiles, being very impressed with how quickly CF-GQN had resumed level flight. I then swivelled my head to look at Jack Keenan. I was shocked at Jack's appearance. He was ashen; he had been so wrapped up in his comic book (his method of dealing with his fear of flying) that he had been completely oblivious to our plan.

I spent the next thirty days apologizing to Jack. Even years later, when he saw me walking toward him he would say, "Here comes that crazy bush pilot." I was engaged in Industrial Electronic Sales at this time and would make sales calls to Jack at CN Telecommunications. We had a good working relationship, apart from Jack's experience with the stall maneuver.

From that day forward, I never conducted an abrupt aerial maneuver unless I was satisfied all my passengers willingly agreed to it. Not long after I had frightened Jack Keenan nearly to death, I was contacted by two ladies from Ottawa. Desme Villiers and Diana Trafford were field workers for the federal government. They were both assigned to carry out some survey with regard to the Indigenous people in the area surrounding Norman Wells. This area included Fort Norman, Fort Franklin, and as far east as Sawmill Bay on Great Bear Lake. The women had different jobs but often needed to travel to the same location, so they would charter my Beaver and save the taxpayer money. They were both easy to get along with, and we spent many hours together without any problems—I would provide transportation and they would get on with their surveys.

One afternoon we were on our way to Sawmill Bay when fog began to form in our path. Looking back toward Fort Franklin (now called Déline) I saw that fog was developing along that path as well. We were approaching Grisly Bear Mountain, near the east shore of Keith Arm on Great Bear Lake, when I spotted a small lake quite near Jupiter Point and decided that was where we must land.

I called Norman Wells on 5680 kHz and was impressed with the response I received from the Air Radio operator. Certainly, he was in no position to aid me, yet his calm manner offered assurance as I negotiated darkness and fog-covered hills. The lake I had chosen to set down on was quite small; its far end was almost invisible in fog. Once down safely, we had no way of reaching shore; I pushed my paddle into lake bottom and found only muskeg. My logbook carries this notation: "Fog—forced landing—slept with two women." That was because we could not get to shore, and I spread out my bedroll on the floor of GQN after removing the rear seats and securing them to my port float.

The following day, we had to press on to Warren Plumber's fishing lodge on the east end of Great Bear Lake, just north of Gunbarrel Inlet. A very heavy wind from the west sprang up, and I was forced to put several extra ropes on CF-GQN to prevent her from breaking loose. During the process of getting extra ropes from the cabin, I had an accident; ice had formed on the steps of the ladder leading to the cabin and caused me to slip. I skinned my shin rather badly. Desme, a nurse, found some iodine and chided me when I winced as she applied that liquid fire to my sore shin.

Next morning, because of high winds and the fact that the DC-6 had been held up by fog earlier, Warren Plumber asked me to make a couple of flights moving some of his guests to Sawmill Bay so they could get back home onboard that airplane. I was glad to help out. I was rewarded by meeting perhaps one of the most famous of Arctic pilots, Ernie Boffa.

Ernie was without doubt one of the most famous pioneer western Arctic pilots. It was a great honor to have met him. He was at Sawmill Bay with probably one of the last Norseman in that entire area. I envied him flying that airplane because the Norseman was, and still is, my favorite airplane of all time. I had only limited time as pilot-in-command on DC-3 CF-CUE and CF-CUG, as I have said. Had my time in that aircraft been greater, I might have named it as my favorite, but the Norseman still holds that distinction with me.

Later in the day, I found I was in need of fuel. Desme, Diana and I flew to Port Radium for 80/87 airplane gasoline. I also had to visit the mine manager. I told the two young fellows on the dock that I would need my middle fuel tank topped off with 80/87 (which has an orange color) and to pump any remaining fuel into the rear tank.

When I returned, they were pumping the last few gallons into our rear tank. I looked at the chamois filter in the funnel into which the fuel was being pumped. I felt a sinking feeling in the pit of my stomach. The fuel in the chamois filter was a very dirty brown, not the rich orange color it should have had. I signed the bill, rather reluctantly, feeling that I had been given bad aviation (or perhaps some other) fuel.

To make matters worse, we took off with the front tank feeding fuel to our engine. There are three belly tanks in the Beaver, and the "bad" fuel was in the center and rear tanks. Front and center tanks hold twenty-nine Imperial gallons each, while the rear tank contains twenty-one gallons. And I was drawing fuel from the tank that contained a pure, known supply of gasoline.

I had flown halfway across McTavish Arm of Great Bear Lake when I noticed that I was on the front tank. I cursed myself for not burning the bad gasoline first. If bad fuel causes engine failure, it is not as serious if you have "good" fuel to switch over to. That front tank had been partially used, and I could see it had perhaps ten minutes of fuel left in it. Now I was forced to switch to the remaining tanks that might hold "bad" gasoline.

Almost immediately, the engine sounded rough. I was too far from Radium to return and too far north of Sawmill Bay to reach that haven. The surface of Great Bear Lake was rough and becoming rougher as the wind swept across some 120 miles of open water—landing on rising waves offered us little hope of survival. The waves would surely capsize my aircraft. Since I really had no choice, we flew on toward Kokeragi Point, at the entrance to

Deerpass Bay. As we pressed on, I felt a shudder ripple through the airframe from time to time, and I kept both left- and right-hand fingers crossed. Finally, we arrived near the shore at Kokeragi Point; from that point on, we would be within easy gliding distance of various lakes. My Pratt & Whitney R-985 engine purred like a kitten. All the worry and discomfort had been a product of an overactive imagination. I don't know why the fuel had such a dark brown color, but it seemed perfectly good 80/87 aviation gasoline, and indeed, that sealed barrel at Port Radium proclaimed that the contents were 80/87. It was a lesson, however: whenever you are not completely certain that fuel is good, use up that suspect fuel first. Save the fuel that you know is good, in case you need it.

Desme Villiers and Diana Trafford had now completed their surveys and departed for Ottawa. I was back to routine flying; they had been good clients and I would miss them both. I never saw Diana again, but Desme returned a year later. She attended dinner at our house before flying north for another season.

Shortly after Desme and Diana departed, Lee Work told me to fly to a point on the Keele River to move a group of men from their camp near some falls to another campsite at Tate Lake, directly south of Fort Norman. We marked the spots on my map, and because it was Lee's day for flying the Otter, I had to make this Beaver pickup. Lee didn't tell me, however, that he had arranged to pick the men up from an island in the river, two miles east of the falls, since he did not care for the fast current near the falls. I flew over the Carcajou Range and Plains of Abraham in the Mackenzie Mountains and landed a few miles west of the Tigonakweine Range on the Keele River.

I taxied up to the riverbank on my left, just down from the main camp, but nobody showed up to help me dock CF-GQN in the very fast current. I assumed that they couldn't hear my engine over the roar of the falls, so I placed my left float against the bank with just enough engine power to hold her there while I tied a rope to the left front float strut and secured it around a tree. Then I shut the engine down. At that point, a woman came down the path to say that I should not be there. She said that Lee Work had told them he had no intention of picking them up at this camp, and so they had moved to an island about two miles downstream.

Once again I started GQN and applied enough engine power, keeping her in place while I untied the rope, and then I taxied up to the base of the falls. Lifting the water rudders I found GQN turning almost "on a dime," and I opened up to takeoff power. I had no idea that a Beaver could become airborne in such a short time; the current in the river below the falls must have been twenty knots. I flew down the river and made the first of two ferry trips, taking a load of gear with two men, then returning to get the balance of their things.

When I returned with a second load, I found their tent in ruins; the men had set up a tent and with nothing else to occupy them had gone fishing in the lake about a half-mile away. While they were gone, a bear had come into camp and ripped the tent to shreds. The next day, a new tent was flown in, and fortunately, no more was seen of the bear.

A few days later, in Fort Franklin, I was approached by the forestry service. Forestry was building a young people's recreation camp on Lac Ste. Therese, about fifty miles south of Jupiter Point.

As I couldn't get aviation gasoline in Fort Franklin, I was somewhat reluctant to attempt two trips, which would be required to transport building materials. I was also leery of how a 1,000-foot hill that lay between Fort Franklin and Lac Ste. Therese would affect fuel mileage. I calculated my fuel consumption with reduced power settings to see if I could operate at twenty-two inches of manifold pressure and 1,500 revolutions per minute. That would allow me to set my power back to approximately 41 percent; I was unsure how Beaver CF-GQN would fly while her engine developed a power output of around 183 horsepower. Certainly, we could make at least one trip and see how our fuel held out. A second trip was probably impossible.

If we could climb over the hill with only 183 horsepower, the second trip might be possible, although I still had to fly back to Norman Walls upon leaving Lac Ste. Therese after the second trip, a distance of 155 miles. We would have to try the first flight to find out. Fuel consumption should be about fourteen gallons per hour at that low power setting. My calculations told me that it would be very close, but we should just be able to make two trips and fly back to Norman Wells with tanks almost empty, providing I could keep the fuel consumption below that fourteen gallons per hour figure.

I found that GQN would climb, albeit very slowly, with only 40 percent power, and we were just able to crawl over the hill; then I was able to throttle back as we descended into Lac Ste. Therese on the other side. Coming back empty, I was able to reduce power even further, and we arrived back at Fort Franklin with a bit more fuel than I had anticipated. A second trip was initiated. Again GQN was able to struggle over that hill with about 100 feet to spare. Most of the flight was done at just more than ninety miles per hour—a remarkable speed, considering the very low power settings I was able to use.

With the second trip out of the way, I had only to make my direct run to Norman Wells. Again, I used only 40 percent power and in due course I arrived at the Wells with less than fifteen minutes of fuel remaining in the tanks. Beaver CF-GQN had done a remarkable job for me once more.

CF-GQN was the fastest Beaver on floats I had ever flown, but in order to achieve 120 miles per hour, of which she was capable, it was necessary to droop the flaps about one and a half degrees; otherwise GQN would attain only 115 miles per hour. Not being an aeronautical engineer, I can't explain how this technique works; but as a pilot, I used this trick quite successfully.

During my stay at Norman Wells in the Mackenzie River Valley, I encountered what was to me, and probably many other pilots, an unfamiliar cloud formation that would sometimes last an entire day. As I flew de Havilland Beavers and Otters out of DOT Lake, where we based our seaplanes, I encountered this unique, magical cloud formation about six times from July 22 until October 4, 1967. A unique nimbostratus would pervade the Mackenzie Valley for many miles, both northwest and southeast, in the area of Norman Wells.

A broken to overcast sky would cover the valley, and great wide tendrils of cloud would extend from the bottom of this low ceiling (based perhaps at 500 feet) all the way to the ground. These were wide columns of cloud, and hundreds of them would support the cloud base that resembled giant tree trunks. Very often, a number of small holes would appear in the cloud base that still covered more than nine-tenths of the officially "overcast" sky.

As we flew out of Norman Wells to our various destinations, we would bank left and right to remain in clear air, while dappled spots of sunshine would light up small areas of the valley below. Between those huge columns of solid cloud, visibility was superb; those little areas would stand out more clearly detailed when lighted by the sun than on cloudless days.

25 Yukon Territory and Home

As time went on, we were kept very busy, often flying from early morning until late at night. On August 15, 1967, we had booked a flight in de Havilland DHC-3 Otter CF-CZO into June Lake, Yukon, for a group of American hunters.

June Lake is above the 5,000-foot level in the Backbone Range of the Mackenzie Mountains, 63 degrees, 30 minutes north and 128 degrees, 44 minutes west, about 140 miles southwest of Norman Wells on a heading of 189 degrees magnetic (magnetic variation of 36 degrees east) which makes it about 225 degrees true.

The Americans were arriving in their own Aero Commander on the morning of August 14. I was to fly them to June Lake in the afternoon. Because there were four hunters in the party and much baggage (heavy clothing, rifles, ammunition and many items which would already be available at June Lake), Otter CF- CZO would have to be used.

We departed DOT Lake at 16:45 hours on August 15; about a mile from the lake, a propeller seal blew and it was necessary to quickly return to our base, with oil spraying over the engine cowling and onto the windscreen. I apologized to the Americans and told them that I was sure our base engineer, Malachi McCarten, would have the problem repaired in time for a morning flight to June Lake.

All their cargo was offloaded as Mai went to work on removing the three-blade Otter propeller so he could install the replacement seal. In the meantime, I carried on with several flights in our de Havilland Beaver, CF-GQN. One of those flights was across to Fort Franklin and another to Fort Norman.

It was nearly 23:00 hours—daylight in the land of the Midnight Sun—when I had finished for the day. Mai had replaced the seal, and he suggested a test flight so we would be all set for our hunters early in the morning. Mai McCarten sat in the copilot's seat for the test flight. Just as we crossed the shoreline, the seal let go again. A hasty retreat put us back on DOT Lake; Mai went to work once again on that pesky prop.

The pilot of the Aero Commander came to our base to lend a hand, and the three of us removed the propeller once again. Since I had to fly early the following morning, I had no choice but to leave them to it.

Our Americans were on deck at 06:00 hours, ready for me to take them to June Lake. Mai had more bad news: that propeller seal had let go once more during a run-up at the dock. There was no way CF-CZO would fly that day.

I must congratulate the American hunters for their patience on that morning. They could have complained bitterly, and in fact one would almost expect it. They understood our dilemma, and we greatly appreciated their patience. "Can you use the Beaver for the trip?" asked one. I said that we really had no choice, but that we had a number of people who had to catch Pacific Western Airlines DC-6, and I would first have to round them up, from a variety of surrounding places, before I could make that June Lake flight.

First I flew to Fort Norman and returned with PWA passengers, after which I had to go to Fort Good Hope. The Canadian National Telecom crew needed to be brought from the Chick Lake repeater to catch the DC-6. Each time I arrived back at Norman Wells on DOT Lake, my Americans would ask. "Can we go now?"

"Not yet," I would have to reply, "I still have a few more people who need to catch PWA."

Finally, after another trip to Fort Good Hope, I had transported everyone who needed to get to Norman Wells so they could depart on that DC-6 Flight. At last, I told the hunters, we were ready for their flight into June Lake. I had already flown 266 miles and transported twenty-three passengers. I had many more miles to go before I would sleep.

My hunting party cooperated with me. The de Havilland Beaver is quite a lot smaller than the Otter. Some items simply had to be left behind. We eventually decided on which items were absolutely needed and placed the balance in our warehouse at our DOT Lake base.

As I said previously, June Lake is around 5,000 feet above sea level; the Beaver's Pratt and Whitney R-985 engine, while supercharged, would need all the power she could develop at that altitude to carry our fully loaded aircraft to the Lake. Because Lee Work and I alternated our flying schedules, I was glad Lee would make the pick-up of these hunters, using

CF-CZO a few days after this trip in Beaver CF-GQN.

Just before we left the dock, Audrey Work arrived with a sandwich and a thermos of coffee for me, since I had already missed lunch.

CF-GQN was more than fully loaded, even after we had removed gear which was not absolutely essential. I didn't know it then, but a good many bets were being placed with Mai McCarten; they bet against my getting GQN airborne with the remaining big load. McCarten had more faith in me than I would have believed. Mai accepted each and every bet; as a result, he made a good deal of money that day. That was the day I decided that the only way to prevent a Beaver from getting off the water would be to sink her.

Because of our load, not only its weight but also its bulk, the center of gravity aft was well beyond its limits; I knew that from the way she wallowed along on the water with her rear spreader bar well submerged. But she responded well to takeoff power, and after a bit of coaxing, CF-GQN was on the step and in the air. Mai McCarten began collecting his winnings.

Our climb was somewhat ponderous, but little by little, as fuel was burned off and we approached the mountain foothills, we began to gain some altitude. Soon we had climbed high enough to clear some 7,000-foot peaks. GQN was unstable; this always happens when the center of gravity is beyond the aft limits. I was forced to hand fly her every minute of the flight. As hungry as I was, I was unable to eat the sandwich or drink that coffee which Audrey had prepared for me.

We always tried to avoid such load imbalances, but there are times when it simply can't be avoided. If we had access to a second Otter, or if time had been available for two trips with the Beaver, we might have been able to surmount the problem.

After about an hour and thirty minutes in the air, we arrived at June Lake. When we had docked, one of my passengers began complaining about his back. The poor guy had been unable to shift his back away from the pressure of a rifle butt—he had suffered constant pain during the entire flight without complaint. After we landed he had a distinct lean to one side as he walked away. Eventually he was able to stand straight. Too much gear had been packed into too little space to allow any movement for passengers.

Upon my return from June Lake, I had a trip to Kelly Lake, just about twenty-six miles northeast of DOT Lake (on the other side of the Norman Range). When I returned, we had only a short trip to the Dredge, at San Sault Rapids. I hoped this would end my long day and I would get some much needed rest.

On August 17, 1967, I flew 1,121 miles. This had certainly proved to be a busy work period at Norman Wells. A few days later, I was called to fly south to a lake near Wrigley. That radio call came in from an Shell Oil Explorations field camp. They were operating out of a lake west of Wrigley and wanted to be moved to a lake just east of Wrigley.

Their helicopter, which was used for local exploration flights, was simply too limited in load capacity to relocate their camp. They requested de Havilland Beaver CF-GQN for the move.

Since I was asked to move their camp, they asked that I pick up a few gallons of engine oil for the helicopter. I arrived at the camp and loaded the men and their gear, taking off for Wrigley; the helicopter pilot took on the load of oil I had brought and other items needed for his operation and then also left the old campsite.

My landing was one of those silk-smooth arrivals one always hopes to execute when there are witnesses. A few minutes later, when the helicopter arrived, we were discussing our landing. I spoke with the helicopter pilot and asked a hypothetical question.

"Suppose," I asked, "that I had hit a rock during that landing and upon your arrival you had a heart attack, what kind of chance, do you think, would I have of flying you in your helicopter to the hospital?" He looked at me for a moment, sizing me up, and said, "I don't think you'd have much of a chance. You would be almost sure to kill us both, or at least destroy my helicopter."

I asked him why he thought that would be the outcome. "Well," he said, "I assume you have never flown a helicopter before, correct?" I said that I had not. "Helicopters have a natural tendency to drop their noses and roll to the right. This happens almost instantly. To prevent it from happening, you have to anticipate, by using left pedal and with a fair amount of stick input, back and to the left."

He looked across at the hills to the east of the lake before he continued. "A certain finite time is needed for any mechanism reaction, and once the machine has begun to dip forward and to the right, you'd find it too late to correct it. The rotor would have already gone into the ground." I asked him, "What if I did what I should do, anticipate, but moved the controls too far to left and to the rear, would she not roll over to the left and dig her tail rotor?" He told me that it takes very little time for the machine to go in the direction it wants to go, so it would only be necessary for me to relax my control input and allow the helicopter to make its own corrections. I was happy to be flying a Beaver, which didn't react so strangely. I suppose by now, twenty-nine years later, helicopter designers have put a fix on those undesirable characteristics. But in the late 1960s, helicopter technology required a very dextrous pilot's hand on the controls.

The oil company crew had come upon an old, unused cabin near their old camp; they had come away with a photograph of the most decrepit de Havilland Beaver I had ever seen. Paint, what little could be seen, was peeling and several very dark splotches could be seen on her fuselage panels. Upon closer inspection, I could see the registration, CF-GQN. I looked at GQN, with her smart red, white and blue paint scheme, sitting not twenty feet away. I marveled at the miraculous transformation after that photograph had been taken.

Someone in the group mentioned that they had long since run dry of spirits and suggested that since there was ready transportation at hand, they should put in an order for a few bottles of whisky, rum and brandy. But a question was raised by another group member about how such a flight could be justified to the bean counters back in Edmonton. In the end, somebody else suggested that surely no one would question the resupply of more helicopter oil.

I flew back to Norman Wells with one of the survey crew members and, using the Northward Aviation jeep, drove him next door to our company house, where the liquor store was situated. In short order we were on our way back to Wrigley with several containers of the "helicopter oil."

One day, Bruce Alcorn, from Northward Aviation's Yellowknife base, stopped in on a charter flight. Bruce appeared shaken and I asked him what had happened. Bruce looked at me and said that he'd just had a terrifying experience at Wrigley. "It was not today that it happened, but it was what I saw today that frightened me." Bruce went on to say that on this past Christmas day, no other airplanes would land at Wrigley, because of poor visibility and low ceilings. Alcorn was approached by several men who wanted to go home to Wrigley that day.

Bruce Alcorn had never been to Wrigley until that Christmas, but he had a hand-drawn detail map of the non-directional beacon and the runway position, so he decided he would take on the charter in spite of the poor weather conditions.

He had navigated to Wrigley on top of cloud, and using the NDB, he found his destination and executed a let down on instruments. The ceiling was only about 200 feet, and he located the runway. He took off heading north—the wind was from the north—and executed a right-hand departure almost immediately. He then set up a climb over the rocks of the McConnell Range. He made one more trip into Wrigley that day.

He had never been to Wrigley before. And nobody had told him of the sugarloaf hills that dotted the valley or about the large rock (about 1,200 feet high) in line with and north of the runway. Unaware of the hazards, he received quite a shock as he landed on the Wrigley strip. He told me later that he was shaken afterwards by the thought of "what might have been."

Two days following the Wrigley episode, I had three men to fly into Little Dal Lake. They wanted to look over some property there and then go on to Trench Lake, about sixty-five miles southeast. We departed Little Dal Lake under an overcast sky that obscured the tops of the surrounding mountains. Not far north of Little Dal Lake, I saw on the map a pass that led from the valley I was flying over to the next valley to the east. If we could fly through that pass, we would find fairly low ground all the way to Trench Lake. Although it would not offer a straight path, because I would have to follow several of those valleys through the mountains.

We spent the night at Trench Lake, and next morning I flew the clients to Fort Simpson. At that point, I was told they had some business at Fort Liard, south of Nahanni Butte, not far from the B.C. border and just east of the Yukon and Northwest Territories boundary. I can't recall the reason for their trip into Fort Liard, but I shall never forget that charming settlement, with its tall, straight flag pole and antennas all painted a sparkling white. The well-kept homes were painted white with a red strip and were surrounded by white picket fences. It was my only trip into Fort Liard, but thirty years later I recall it vividly. I left the men at Fort Liard and returned to Norman Wells, via Low Level Airways Amber Route Six.

I had learned a good deal about mountain flying from John Davids; his knowledge helped me stay alive. I now turned to other flying chores as the season began to wind down. It was then that I met a gentleman who was living proof that the North is not only for young people.

His name was Hammer, although I never knew if that was his first name or his last. Hammer was getting on in years; he must have been about seventy-five at the time. One day he showed up at our Norman Wells Seaplane Base, wanting a trip into the hinterland of the Northwest Territories. I never knew if he was a prospector, hunter, trapper or just a man who loved the wilderness.

As we loaded his belongings on board de Havilland Beaver CF-GQN I was surprised at how little bush gear he was taking along. There was a window frame, a wooden door made of narrow boards with the classic angle brace across its back, a hammer and a box of nails and a frame saw, which was folded down so as to take up smaller space, with its blade packed away with other goods. Additionally, he had an ax, cooking pots, a large bag of dried white beans and a large slab of pork—he'd certainly have good pork and beans for his meals. Presumably, because of his 30-30 rifle, he would provide himself with fresh meat.

I found myself drawn to this elderly outdoorsman, partly because of his very young outlook on life.

"You see," he told me, "I use only iron cooking pots. Last year I had only aluminum ones and became very sick around Christmas." Hammer stopped to set a match to his pipe, packed with Bond tobacco, and went on. "I am sure the aluminum will come off the pots and that will make people sick."

I had been told previously that he staggered into Norman Wells on Christmas day the previous year; he'd had to walk over 120 miles, sick as he was. He managed to make fairly good time; I believe he was flown to the Inuvik Hospital, where he had regained his health. At the time, it seemed, nobody put much faith in Hammer's theory. Now there appears to be some evidence to lend support to his belief; aluminum cooking utensils may not be good for one's health. I have even heard that there may be evidence to suggest that aluminum might play some part in promoting Alzheimer's disease.

Hammer's cargo, of course, contained other foodstuffs, an assortment of hand tools, such as screwdrivers, chisels and cutlery, and of course a coffee pot, plus his collapsible canoe for summer transport. Hidden within his pack, I spotted a pair of snowshoes for winter travel.

This flight took place twenty-nine years ago on a September day, but I believe I flew Hammer to Fifty Mile Lake on that occasion. I do recall that I offloaded Hammer and his load on a small point of land. I was concerned about Hammer's well-being, since he would be out on this isolated lake for several months. If he found himself with a need for medical help, I could see no way for him to seek aid but to walk out (as he had indeed done a year previously). He had no radio. In spring, walking out through muskeg, crossing creeks and skirting lakes would have been difficult, if not impossible, should he need to reach medical aid. In winter, he would have been more mobile, as he could strap on his snowshoes.

"Will you be okay out here all alone?" I asked. He had a somewhat interesting accent that I was not able to identify; it could have been Norwegian, Dutch or German, but it seemed as though it was none of them, or perhaps it was a combination of all of them.

"Well," he said, "I'm always okay, when I am out all alone in the bush; it is when I'm most happy."

I left Hammer on his generously wooded peninsula and taxied out into the little bay in front of his campsite. He was a better man than I was. I'd not be able to allow my only means of transportation to disappear into the sky.

"Write in the book to pick me up December 18," he said.

I would not be on that pick-up, because by December 18 I would be long gone from the Territories; I really and truly hope that somebody made that pick-up.

As the days dwindled down toward freeze-up, I would deviate from my course while en route to some other destination in Hammer's direction and drop in to see how he was doing. I was amazed. Hammer had felled trees and had built himself a cozy log cabin, overlooking the little bay where a small island stood about 100 yards offshore, containing a stand of four or five trees. Moose would often swim to feed on the lush grasses growing on the island.

Hammer was the soul of hospitality. He offered me a cup of coffee and sometimes a plate of his delicious pork and beans. During one of my several visits, Hammer had shot a deer. It was clear that he was to have meat on his table for some time to come.

Over the past twenty-nine years, I have often thought of Hammer, and every time my thoughts have turned in his direction, I remember him as a soft-spoken man of the northern bush country. Of course, it is likely that he has gone to his reward by now. But in my mind, I see him now as I last saw him.

Fall was approaching, and darkness came earlier every day as the earth continued in its orbit around the sun. One day, DC-3 CF-CUG—which now belonged to the Department of Transport—visited Norman Wells on an annual inspection tour of various airports in the Territories. By this time, Malachi McCarten had moved on to another base and Bill McKinnon had replaced him. Bill told me several times that his father used to say, "I'd rather spend eternity in Hell with a broken back than a single night in Wabowden, Manitoba!"

That evening, I was returning from a trip to Fort Franklin and I found the night was rapidly approaching; I would have no choice but to land illegally in the dark. I approached from the south in order to keep my Pratt & Whitney R-985 from alerting those at the airport. This would be, essentially, a glassy water landing, since I could only see the outline of the lake surrounded by darkness of the trees. I misjudged slightly and came over the shoreline just a bit too high. I was sure I could not land and get turned around before running out of lake, so I executed an overshoot. Of course, my Beaver produced plenty of noise in that still, evening air. By the time I had come around and executed a proper landing, anybody who was interested would know that I had arrived. I taxied to the dock, and as I tied up. Bill McKinnon came tearing up with the company jeep.

"Man." Bill said, "are you in deep khaki-poo or what?"

"What do you mean?" I asked.

"Didn't you know the DOT guys are up at the airport?" Bill asked.

"Well, yes," I replied, "but surely they have better things to do than worry about one airplane arriving a bit after dark."

"Oh boy!" Bill said. "I'm glad I'm not you."

In a few minutes I was back in our company house. I began to think that Bill might very well be right. I waited for nearly an hour to hear from the inspectors. When nothing happened, I telephoned the Air Radio office.

When the operator on duty heard my question, he began to laugh.

"Oh, that Bill McKinnon again." He laughed. "He's always trying to put something over on pilots. No. The inspectors are probably over at the Gulp and Puke having a couple of drinks. I heard your engine okay, but I'm sure they would never hear anything in that noisy place."

I relaxed, and when I cornered Bill, he just laughed; he did have a history of trying to upset pilots.

Glassy water landings in any de Havilland Beaver were unique in that the engine noise would radiate downward from the airplane and reflect back up to the aircraft structure. This reflection would produce a very special sound, a sort of hollow, drumlike rumble. The nearer to the surface the keels of the floats approached, the more hollow the sound would become. After executing enough glassy water landings, I got to the point where I could judge within a few inches how close to the surface my floats had come. This was only useful on glassy water; water with ripples or waves seemed not to reflect in the same way as did glassy water.

Just a few days following my last visit with Hammer, pilots began flying south for freeze-up. One pilot flew in from Inuvik on his way to Edmonton. His young husky dog wandered off, and when his Beech 18 aircraft was ready to leave the dog was nowhere to be seen. "If you find him," the pilot said, "bring 'im along with you to Edmonton, will you?" The man gave me a telephone number in Edmonton where he could be reached, whenever I arrived with his young dog. Within half an hour, that young pup showed up. He took to me right away. I have always been a cat lover and have had very little to do with dogs, but I had promised to look after this guy for the other pilot, so I took him back to our company house. We got to know one another a bit.

He was a funny little fellow and liked to nip at my toes when I had come home and removed my boots.

By this time, the weather was beginning to turn a bit colder. Each morning we had a thin crust of ice formed during the night, clinging around the edge of our dock on DOT Lake. Summer flying season was almost behind us, and it was time for me to take CF-GQN to Cooking Lake (just east of Edmonton).

I had begun growing a beard while I flew the Petro-Par people around; I suppose that beard and the dog made me look the part of a grizzled old Arctic pilot. Since we would be parting company before long, I never bothered to give the husky a name; I called him, instead, just plain "Dog" and we became quite good friends. Dog and I departed Norman Wells on September 1, 1967.

I placed my map case between the flight-deck seats, and Dog would climb up on top of that case with his chin resting on my right thigh, while I draped my coat with its imitation sheepskin lining over him. He would sit there, apparently in total contentment, for hours as we flew southeastward, above the Mighty Mackenzie, rolling its mile-wide tide along towards the Arctic Ocean.

Because of low clouds, I elected to climb on top of the cloud deck. After all, I was again following that Airways Amber Route Eight, and we were out in brilliant sunshine. During that flight, I followed Amber Route Eight just west of the Mackenzie River, in the vicinity of McGern Island, nearly straight east of Little Dal Lake, where Bob Shepherd was killed coming out of that lake. If Bob had only flown a couple of miles north, to the next opening ...

Nothing short of a stalled turn would have saved Bob's life in that box canyon. I used to practice stalled turns about every two or three weeks so I would be sharp enough to carry it off if the need should ever arise.

One of the fellows who found Bob and his Beaver told me later that his floats had scratched along on the eastern wall of the canyon, leaving a well-defined set of float marks, and Bob's aircraft had crashed into the north wall. When I was told about this, I felt sad. Having seen him nervous while flying, I was not very surprised. Bob Shepherd was a good man; it is a shame that we lost him.

Dog and I flew beyond Fort Simpson, pressing on to Hay River. I hadn't been to Hay River for some time, but I remembered the sign I had seen at the local restaurant: "No pets." What could I do with Dog, I wondered? I couldn't just turn him loose in the street; he would become lost again. I decided that he was small enough still to hide under my big coat. One of the waitresses spotted him, and she was delighted with him. She went into the kitchen and fetched a nice piece of steak in a tin dish; we placed it on the floor by my foot and I slipped Dog down there to keep prying eyes from seeing him. The weather turned sour on us, and we had to stay put for two days. Each mealtime, Dog had his nutritional needs met by a waitress in the cafe. It was not always the same one, because they all thought Dog was so cute.

When the weather had cleared sufficiently, we continued on our way to Fort McMurray, flying over Fort Chipewyan. For the first time, I noticed the expanse of brilliant white sand in the area. I was beginning to feel a cold coming on, and the nearer we got to Fort McMurray, the worse I felt. Clarence and Norma Palsky asked me to stay with them, because the weather had also turned bad on us again.

I was not the only southbound pilot stranded at Fort McMurray. Several others from Gateway Aviation, including Cedric Mah, became weather-bound as well. I was able to meet with Cedric and discuss, at length, our experiences with Pat Parker, who had such a lack of bush and Arctic sense.

The Palsky children, particularly the youngest member of that family, Bruce, came to love Dog. Since Norma was a nurse, within three days, her care helped me beat my cold. The weather began improving, and soon I was able to complete my return to Cooking Lake. When I arrived at Cooking Lake I telephoned the pilot to whom I had made the promise about finding Dog and bringing him back. The man asked me to order a taxi to deliver Dog to his master. We had become very attached to each other, and it turned out to be difficult to say goodbye and watch Dog depart in that taxi.

I was later told that Dog, being a husky, as an adult would require a deer or half a moose every day—an exaggeration, no doubt, but the point was well taken.

The de Havilland Beaver CF-GQN had served me well. We had flown many miles and done many things together, and I would miss her sorely. My logbook records, "Sad day ... Had to say goodbye to GQN and to Dog." I had enjoyed my bush and Arctic flying experiences, and I had been well rewarded financially for my efforts. But I never did fly for money; I flew for the love and freedom of flight. In those days, the bush and Arctic pilot was probably the last totally free human on the planet. That was worth all the financial rewards anyone could offer.

Those days are gone forever. Freedom no longer is part of flying in the Territories. Air traffic control has reached the outermost regions of the Arctic. Global Positioning Systems have replaced the pilot's self-reliance. My commercial flying career was over; I would fly again, but not as a professional pilot; my days of bush and Arctic flying had come to an end.

Bush and Arctic Pilot

Epilogue

I retired from commercial flying when I arrived back at Cooking Lake because our son Mark was nine years old and Julie and Kevin, our twins, were six. I felt it was important to be around for them. Based on the quality of citizens they have become, I feel this was a wise decision. Many bush and Arctic pilots with young families have continued their careers, and I sometimes wish I had done the same. But I am convinced I followed the right course.

I joined Canadian Electronics Limited as an outdoor electronics sales representative, often utilizing a Piper Twin Comanche CF-USF to visit customers many miles from Edmonton. In the spring of 1971, I accepted a position with Cardinal Industrial Electronics, also as an outdoor sales engineer. Bill Bissonnette was a partner of that company and has owned several aircraft, which we used for business and personal transport. We used a Cessna 180 CF-HCW, Piper 140 CF-URZ and Cessna Cardinal CF-DJL.

During those years, from 1967 to my retirement in 1992, I flew a variety of aircraft, which even included a Baby Great Lakes trainer CF-RLK that was hand-built by Raymond Kloosterman.

I am now retired from both aviation and the electronics field and live in Edmonton with my family. I shall always prize my flying experiences, and I have established lasting friendships from those bygone days.

RELATED TITLES

Then & Now

This account, of the geography and history of some of the mountainous country drained by the South Nahanni River, is based on Lougheed's observations as a hiker and paddler, and on her thorough research — including interviews and correspondence with the people, and the descendants of the people, who made that history..

Nahanni
Lougheed, Vivien

978-0-88839-697-6 [paperback]
978-0-88839-697-6 [epub]
5½ x 8½, sc, 246pp

$24.95

Burnt Snow
Moore, Kieran

978-0-88839-100-1 [paperback]
978-0-88839-356-2 [hardcover]
978-0-88839-265-7 [epub]
6 x 9, sc, 272pp

$24.95

My years living & working with the Dene of the Northwest Territories

The reflections of the authors encounters with some of the leading figures of the North are quite humorous and consequential in the later development of the North. He describes the Indigenous Elders who would influence him in countless ways, and how their teachings are later, the source of northern survival in otherwise seemingly impossible situations. This book reflects the people of that time, and their lifestyle of living off the land in total independence and their incredible life-skills of survival.

27 Years Off-grid in a Wilderness Valley

The Power of Dreams tells the story of a couple, already in their 40's, who uprooted themselves from urban life to follow their dream of living in the wilderness. They settled in a remote mountain valley called Precipice Valley, part of the ancient trade route linking B.C.'s Chilcotin plateau to the Pacific Coast. Surrounded by mountain vastness they lived there for nearly three decades, much of it in near-total isolation. Their dreams sustained them while they carved out a lifestyle that was both rewarding and challenging.

The Power of Dreams
Dave & Rosemary Neads

978-0-88839-718-8 [paperback]
978-0-88839-742-3 [epub]
5½ x 8½, sc, 246pp

$24.95

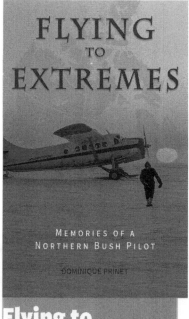

Memories of a Northern Bush Pilot

Recalling some of the most memorable escapades ever conducted in the Canadian Arctic with bush planes, *Flying to Extremes* takes place in the late '60s and early '70s from a base at Yellowknife, in the heart of the Northwest Territories.

This entertaining book recollects Prinet's adventures as a young man while capturing the humour, beauty, danger and unique culture of northern communities, in the dramatic landscape of the Canadian Arctic. Readers familiar with the region and those who can only dream of visiting it will both find this title a nostalgic and captivating read.

Flying to Extremes
Dominique Prinet

978-0-88839-145- [paperback]
978-0-88839-234-3 [epub]
5½ x 8½, sc, 280pp

$24.95

Hancock House Publishers
9313 Zero Ave, Surrey, BC V3Z 9R9
www.hancockhouse.com
info@hancockhouse.com
1-800-938-1114

OTHER AVIATION TITLES FROM HANCOCK HOUSE PUBLISHERS

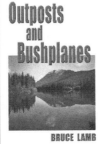

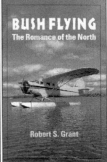

Broken Arrow #1
John Clearwater
978-0-88839-596-2
5½ x 8½, sc, 160 pp

Outposts & Bushplanes
Bruce Lamb
978-0-888395-566
5½ x 8½, sc, 208 pp

Packtrains & Airplanes
Trudy Turner
978-088839-710-2
5½ x 8½, sc, 424 pp

Bush Flying
Robert Grant
978-0-888393-500
5½ x 8½, sc, 286 pp

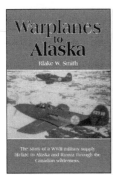

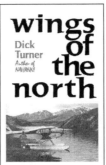
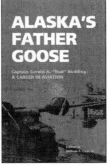

Warplanes to Alaska
Blake Smith
978-0-888394-019
8½ x 11, sc, 296 pp

Wings Over the Wilderness
Blake Smith
978-0-888395-955
8½ x 11, sc, 296 pp

Wings of the North
Dick Turner
978-0-919654-617
5½ x 8½, sc, 288 pp

Alaska's Father Goose
Gerald Bodding
978-0-88839-651-8
5½ x 8½, sc, 176 pp

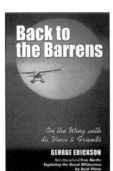

Back to the Barrens
George Erickson
978-0-88839-642-6
5½ x 8½, sc, 328 pp

Hancock House Publishers
19313 Zero Ave, Surrey, BC V3Z 9R9
www.hancockhouse.com
info@hancockhouse.com
1-800-938-1114

Made in the USA
Middletown, DE
05 March 2024

50224060R00144